The Frontiers Collection

The books in this collection are devoted to challenging and open problems at the forefront of modern science and scholarship, including related philosophical debates. In contrast to typical research monographs, however, they strive to present their topics in a manner accessible also to scientifically literate non-specialists wishing to gain insight into the deeper implications and fascinating questions involved. Taken as a whole, the series reflects the need for a fundamental and interdisciplinary approach to modern science and research. Furthermore, it is intended to encourage active academics in all fields to ponder over important and perhaps controversial issues beyond their own speciality. Extending from quantum physics and relativity to entropy, consciousness, language and complex systems—the Frontiers Collection will inspire readers to push back the frontiers of their own knowledge.

Liam Graham

Physics Fixes All
the Facts

 Springer

Liam Graham (iD)
London, UK

ISSN 1612-3018   ISSN 2197-6619   (electronic)
The Frontiers Collection
ISBN 978-3-031-69287-1   ISBN 978-3-031-69288-8   (eBook)
https://doi.org/10.1007/978-3-031-69288-8

This Springer imprint is published by the registered company Springer Nature Switzerland AG
The registered company address is: Gewerbestrasse 11, 6330 Cham, Switzerland

If disposing of this product, please recycle the paper.

Advance Praise for This Book

"I wish I knew as much as Liam Graham. It would have enabled me to write a much more convincing and well informed book than *The Atheist's Guide to Reality*. Fortunately Graham has done it. My envy of Graham's erudition is only surpassed by my admiration of his achievement. This is the definitive guide to why the physical facts fix all the facts! It's also the definitive diagnosis of all the specious arguments against this simple truth."

—**Alex Rosenberg**, R. Taylor Cole Professor of Philosophy, Duke University, author of *The Atheist's Guide to Reality*

"Graham does an impressive job of advancing his vision of 'austere physicalism' as against non-reductionist views on which there are higher-level or 'emergent' phenomena. His lively, wide-ranging, detailed treatment of the relevant scientific case studies and philosophical positions is a tour-de-force, and his critical salvos and defensive strategies deserve further attention by scientists and philosophers alike. Anyone curious about the structure of natural reality will find this book to be a great read and a valuable resource."

—**Jessica Wilson**, Professor of Philosophy, University of Toronto, author of *Metaphysical Emergence*

"This well written book offers a balanced approach for those with interests in physics and/or metaphysics. It dismisses various forms of emergentism, arguing that these views wrongly project human cognitive limitations onto the world's ontology. It defends an austerely monistic version of physicalism according to which the world is a single entity—the 'blobject'—with richly complex dynamic structure but without any constituent entities as proper parts. Diverse metaphysical topics are addressed, including free will and consciousness. Highly recommended."

—**Matjaž Potrč**, Professor of Philosophy, Ljubljana University, author of
Austere Realism

"Whatever you think about the nature of reality, there's value in grappling with the idea that it may fundamentally be 'all physics'! Liam Graham presents an engaging and well-researched argument, with some excellent examples drawn from across the sciences."

—**Louis Barson**, Director of Science, Innovation and Skills, Institute of
Physics

"This book is a thorough and critical examination of the idea of emergence arguing that the concept is so generic that it is useless. It provides a very good overview of emergent phenomena, particularly those from condensed matter physics, and is written in an entertaining, thought-provoking style."

—**Ilias Amanatidis**, Ben-Gurion University of the Negev, Israel and
Ioannis Kleftogiannis, National Center for Theoretical Sciences, Taiwan

"I have often puzzled over claims that emergent properties are 'something else, something that cannot be explained by the elements of the system'. This splendid book shows why such claims are nonsense. And it helps us understand why, in a few years, that thesis will not be in the least controversial."

—**Antonio Cabrales**, Professor of Economics, Universidad Carlos III,
Madrid

For Axelle

Preface

Books in the humanities often begin with a statement of the author's position. Science books rarely do. Since this book involves as much philosophy as science, let me start by describing where I am coming from.

I have a long-standing dislike of mystical or magical thinking in all its forms. A dislike of thinking that avoids rigorously seeking a good explanation and opts instead for an attractive one. Of thinking that settles for a baroque explanation rather than accepting that some things are as yet unexplained. The usual suspects of free will and consciousness are fertile ground for such thinking, as is emergence, a term widely used to describe complex systems and a central topic of this book. This means I am an opinionated narrator. But I strive to be a reliable one and include extensive references and suggestions for further reading to help you make up your own mind.

Let me give an example of what motivates me. Later in the book I will cite a philosopher of science who argues that the placebo effect is evidence against physical causal closure. I find this deeply suspicious. Causal closure is right down at the fundamental level of quantum physics. The placebo effect, while well documented, is a property of the human brain, the most complex and poorly understood system we've come across. No evidence is given which links the two. Perhaps the philosopher will turn out to be right. But for now there is no reason to think that our lack of understanding of the brain should have any implications for physics.

Descriptions of emergent phenomena often convey little more than "Wow, that's so mind-blowingly complex it can't be just physics". For those who want to see more clearly, this book shows how emergence can be eliminated and presents an unflinching worldview in which everything, without exception, is physics.

London, UK   Liam Graham

Acknowledgements

Many thanks to my editor Angela Lahee and the series editor Rudy Vaas for their enthusiasm, feedback and support. A big thank you to Divya Sureshkumar and Sangeetha G Ganesan for guiding the book through production.

My thinking has been shaped over the years in discussions with friends and colleagues too numerous to mention. Let me thank those who shared their thoughts on this text: Ilias Amanatidis, Sue Arthur, Louis Barson, Alex Buell, Antonio Cabrales, Nigel Goldenfeld, Ioannis Kleftogiannis, Honor Klein, Paul Mans, Kanesh Rajani, Jessica Wilson and Stephen Wright. Special thanks to Nick Rimmer for his intuition, energy and remarkable attention to detail.

Contents

About the Author

Liam Graham "Do we need more than physics to understand the world?" Liam first asked himself this question as a teenager and it has been the driving force behind his career ever since. After a degree in Theoretical Physics at Cambridge and a master's in Philosophy at Warwick, he eventually found economics to be an appealing middle ground and completed a Ph.D. at Birkbeck College, London. To pay the rent, he taught English, developed and sold trading software and was the numbers' guru for a boutique finance house.

Liam's 15-year academic career was mostly spent as an Associate Professor at University College London, working in one of Europe's top Economics departments. His research involved building mathematical models of an extremely complex system, the macroeconomy, and his work was published in all the top macroeconomics journals. Whether working on philosophy or economics, he never stopped reading science and exchanging with scientists. In 2018, he left UCL to concentrate on his original question and the wide-ranging, multidisciplinary and endlessly fascinating project it has become. His first book, *Molecular Storms: The Physics of Stars, Cells and the Origin of Life* was published by Springer Nature in 2023.

1

Introduction

Hurricanes. Living cells. Flocks of birds. You yourself. Few would deny that these things are made of atoms. Yet they behave very differently from atoms. Fundamental physics might do a good job of explaining atoms, but such complex phenomena seem to lie outside its scope. This is the basic idea of emergence. Things emerge from physics but are beyond physics. The whole is greater than the sum of its parts. More is different.

Emergence is one way of understanding complexity. There are alternatives. You can be a dualist. Then some things are supernatural, in a different domain from physics. More is spooky. This might seem to apply only to the last item on my list, but it wasn't so long ago that hurricanes were seen as avenging ghosts and life as caused by a vital spirit.

Or you can be a physicalist. In this case, everything is physics. More may be different but more is always different. Physics explains the properties of the whole and the properties of the parts. The nature of quantum physics means the whole can influence the parts as well as the parts influencing the whole. If we don't fully understood things, this is a result of lack of knowledge or computing power.

Many find neither alternative attractive. Emergence promises a middle way. You can have your cake of not believing in the supernatural. And eat it with the pleasure of knowing that, while it is made of atoms, it is somehow more than those atoms. Stephen Hawking said in an interview:

© The Author(s), under exclusive license to Springer Nature Switzerland AG 2025
L. Graham, *Physics Fixes All the Facts*, The Frontiers Collection,
https://doi.org/10.1007/978-3-031-69288-8_1

The human race is just a chemical scum on a moderate-sized planet, orbiting around a very average star in the outer suburb of one among a hundred billion galaxies.[1]

Emergence allows us to accept we are a chemical scum while rejoicing in being more than just a chemical scum. This appeal is part of the reason for the remarkable spread of the term. It can be found everywhere, from fundamental physics to chemistry and biology, to sociology and economics.

However appealing, it is an illusion. Emergence is usually divided into two types according to its relation to physics. Weak emergence is consistent with current physics. Strong emergence is outside current physics. This book will argue that neither tells us anything useful about the world. Weak emergence turns out to be so weak that it can be applied to everything. And strong emergence is such a strong criterion that there is no evidence for it. The term emergence either refers to everything or to nothing. We think it tells us something about the nature of reality, but this is an illusion.

Fascinating phenomena exist at every scale but describing them as emergent adds nothing. Emergent behaviour. Emergent organisation. Emergent structure. Whenever you see the word you can simply discard it. You can discard its aura of mystery and its suggestion that some things will be forever beyond our understanding.

If you want to avoid the supernatural, you are left with physicalism. This book argues that the only possible physicalism is an austere physicalism that dissolves our commonsense understanding of the world. Physics fixes all the facts. Any description of the world that is not fundamental physics is at best an approximation. Such descriptions may be useful, they may be necessary but they are functions of our interests and our cognitive structure not properties of the world. This goes for everything that is not fundamental physics: the concepts which make thought possible, our intuitive notions and the rest of the sciences.

These things are illusions. Reality means having causal power. If everything is physics, only the entities of fundamental physics have causal power. Other things are therefore unreal, illusions. So there are no objects. No creatures, colours or concepts. Instead, there are arrangements of quantum fields.

Our sensory limitations mean we can't see quantum fields. Our cognitive limitations mean we can't intuitively understand them. Yet despite these limitations we perceive a world full of structure and meaning. From a physicalist perspective, this leads to fascinating questions. Why do we perceive creatures, colours and concepts? Why does a quantum field arranged in a particular

---

[1] Stephen Hawking, interviewed by Ken Campbell in Dugan (1995), 50′00″.

way interact with another arranged as a human brain so that it adopts a state which corresponds to creature, colour or concept?

More specifically, a physicalist approach allows us to unpack the term emergence and show how it lumps together disparate ideas about the limits of our thought. Emergence may be no more than an assertion that there are interesting questions at every scale. It may be a way of describing phenomena yet to be explained. Or it may be about the distinction between understanding and prediction. Sometimes its use is a result of projecting our cognitive limitations onto the world. Sometimes a result of a failure to distinguish between the nature of reality and the language, models and approximations that scientists use.

Let me now turn to the structure of the book. To start, Chap. 2 gives a broad overview of the sort of phenomena that can be described as emergent. The examples are chosen to cover a wide range of scales and sciences, starting inside the nucleus of an atom and working up through chemistry and biology to mental causation and its place in the universe. These will help illustrate the subsequent arguments and also give an excuse for a romp through some of the most fascinating parts of physics.

Part I presents three general frameworks which will be used throughout the book. Chapter 3 turns to philosophy and identifies six positions: dualism, weak emergence, strong emergence and three varieties of physicalism. Each of these can be understood in terms of where what matters happens. For physicalism, everything that matters happens at the level of fundamental physics. All causation is at the lowest level. For dualism and strong emergence, on the other hand, the system as a whole is what matters. There is downward causation from the whole to the parts and this must contradict physics. Weak emergence describes a precarious middle ground where downward causation is somehow consistent with physics.

One thread of my argument is that the concept of emergence is a consequence of our cognitive limitations, so Chap. 4 describes aspects of human cognitive evolution. Partly this is about understanding our commonsense models of the world, partly about understanding how we transcend them. How can brains that evolved to survive and thrive on the African savannah roam from quarks to quasars? How do they create and take part in the system of distributed cognition which is science? Chapter 5 turns to role of simulations in science and the theory of computation. There are many links to the discussion of emergence and physicalism. The most interesting is the way that quantum computers will radically transform how we simulate systems from the bottom up. Our ability to simulate and hence to understand physical systems may only be limited by the size of the quantum computers we can build.

Part II contains the case against emergence. Chapters 6–9 discuss four forms of weak emergence. The central argument of each of these chapters is the same: weak emergence applies to every real system. If all physical systems can be called weakly emergent, the definitions are empty and the term redundant. Studying these forms of emergence tells us little about the world but much about our cognitive structure.

Underlying all four senses of weak emergence is a basic confusion between the nature of the models scientists use and the nature of reality. Chapter 10 addresses this in the context of three common modelling strategies: the thermodynamic limit, effective theories and the renormalization group. All of these have features which fit one or more of the definitions of emergence. But this tells us nothing about the world, only about the models we use to explain the world.

Next, strong emergence. Chapter 11 discusses possible mechanisms ranging from quantum physics to non-computability. All are logically possible. But there is not a shred of convincing evidence for any of them. Believing in strong emergence is equivalent to believing there are pixies in your garden. Impossible to disprove, but not worth spending your time on until there's some solid evidence. Even if there were such evidence, it would support either an extended physicalism or dualism. As a term, strong emergence is also redundant.

Chapter 12 is a brief summary of the previous chapters. For each type of emergence, it gives a one line answer to three questions: what it is; why it applies to everything and why it is not a challenge to physicalism.

So much for emergence. Part III turns to the alternative. Chapter 13 presents the argument for austere physicalism. It is an easy position to state, but one that some may find self-evidently absurd and most of the chapter is spent dealing with potential objections. Chapter 14 applies it to emergence. If you've discarded the word emergence, what can you replace it with? What becomes of our intuition that more is different?

To wrap up, Chap. 15 returns to the examples, describing them without using the concept of emergence and showing that none represent challenges to physicalism. The chapter ends by throwing down a gauntlet. If you think you have a system which is emergent in the sense that it cannot be explained by physics, there is a simple procedure you can follow to convince a hard-nosed physicalist of your case.

Throughout, I do my best to avoid discussing free will and consciousness. Partly this is because they merit a book of their own. Partly it is a rhetorical choice. If you concede that they are the only place left for emergence, I will consider my job done. But in the Epilogue, I show that there is no reason to think that they cannot be given a physicalist explanation. The chapter ends by revealing the meaning of life.

This book is part of a larger project to investigate the limits of physics. My first book[2] explored thermodynamics and its application to questions ranging from the formation of stars to the inner workings of cells to the origin of life. It concluded that there is no reason why physics shouldn't one day explain all of this. My next book will take the same approach to cognition and consciousness, starting with the simplest systems and working up through cognitive evolution to human subjective experience.

Emergence claims to put some things beyond physics. Addressing this claim is central to the physicalist project. The past decade has seen a dozen or so monographs and collections about emergence. Apart from the odd article, they are all resolutely supportive. This book aims to redress the balance by showing that emergence is an empty concept and providing an alternative framework with which to understand the world.

Humanity starts in a world of incomprehension. Magic and deities are everywhere. The scientific project chips away at this. Replacing intuitions with scientific concepts. Gradually withdrawing magic from the world. Emergence is a last refuge from this process. It promises to rescue the world from the austerity of physicalism. It puts humans and the concepts we use right at the heart of everything. It allows mind, consciousness and humanity to retain something of their previous dignity. It's not so much that more is different, but that I'm different and I know I'm different.

All this is an artefact of our cognitive limitations, an arbitrary way of slicing up the complex physical reality in which we exist, physical systems among others. Emergence is pessimistic and projects our limitations onto the world. Austere physicalism is modest and profoundly optimistic. There are unanswered questions everywhere. But the system of distributed cognition that is science transcends individual cognitive limits. There is no reason to think that we, and the machines we build, shouldn't continue to give us answers.

References

Dugan D (1995) Reality on the Rocks, Part 3—Beyond Our Ken, UK Channel 4.
Graham L (2023) Molecular storms: the physics of stars, cells and the origin of life. Springer, Cham, Switzerland. https://doi.org/10.1007/978-3-031-38681-7

---

[2] Graham (2023).

2

More Seems Different

Summary This chapter introduces the concept of emergence using a broad range of examples. These start from inside the atomic nucleus and work up through chemistry and biology to evolution and mind. While exploring these examples, many concepts that will play an important role in the remainder of the book make their first appearance.

What is emergence? One way of answering this question is by giving examples of physical systems which can be described as emergent. This chapter presents fifteen such examples, chosen to give a broad sweep from the smallest to the largest and across different sciences. There is no shortage of candidates, I could easily have included ten times as many. This means that it is likely your favourite example will not be here.

As a working definition of emergence, let's use the one we've already seen in the introduction: more is different. It comes from a 1972 paper[1] by Philip Anderson (Nobel Prize for Physics, 1977) which is often credited with reintroducing the term emergence into the mainstream. The definition is about composition. Emergence is when the properties of the whole are different from the properties of the parts. It also implies that you cannot understand the behaviour of the parts without understanding the behaviour of the whole.

---

[1] Anderson (1972).

© The Author(s), under exclusive license to Springer Nature
Switzerland AG 2025
L. Graham, *Physics Fixes All the Facts*, The Frontiers Collection,
https://doi.org/10.1007/978-3-031-69288-8_2

A toy example illustrates this. Take some Lego pieces and build a car. This car is emergent. It has properties that the parts don't have: looking like a car, capable of rolling in straight lines or turning corners. More is different. By itself, a single piece of Lego cannot move in a straight line, suspended a few centimetres above the ground. But that's exactly what it does when it's part of the car. To understand the motion of one of the parts, you need to understand the car as a whole.

For the moment, more is different will do as a rough and ready definition of emergence. While working through the examples, I will bring out other senses of the term. These are summarised in the final section and will be discussed in subsequent chapters.

My descriptions of the examples are brief, no more than a handful of paragraphs for each one. For some, there won't be enough physics. If this describes you, the material I present is standard and you can find more in-depth treatments in textbooks or in the suggestions for further reading at the end of the chapter. For others, there will be too much physics. In this case, I suggest you start this chapter in the middle, with the section "Ordinary objects".

One of the aims of this book is to show that the concept of emergence is redundant. So in Chap. 15, I return to these examples and show how they can be understood in a physicalist framework without a mention of emergence.

2.1   Protons and Neutrons

Let's start right down at the bottom, inside the atomic nucleus. While this is a natural place to begin, it involves some of the most complex physics discussed in the whole book. I invite readers unfamiliar with these ideas to skip this and the next couple of examples.

Atomic nuclei are composed of protons and neutrons, called collectively *nucleons*. These are not fundamental particles but are made of quarks. A proton is made of two up quarks and one down quark. A neutron is made of one up quark and two down quarks. An up quark has a positive charge equal to two thirds the charge of an electron. A down quark has a negative charge of one-third the charge of an electron. Combined, these give the charge of a proton equal and opposite to that of an electron and the zero charge of the neutron.

Forces are mediated by particles. The electromagnetic force between charged particles is carried by photons. When two electrons approach each other, the electronic repulsion between their negative charges occurs via the exchange of a photon. The theory describing this is Quantum Electrody-namics (QED).

A further fundamental force is the strong interaction, described by Quantum Chromodynamics (QCD). Quarks experience both electromag-netism and the strong interaction. The equivalent of electronic charge for the strong interaction is known as color. The force is carried by gluons which are electrically neutral but have color so are themselves subject to the strong interaction. This is an important difference from electromagnetism. Photons, the carriers of the electromagnetic force, have no charge so are not affected by the force. When two charged particles interact, they exchange a photon and that is the end of the story.

Things are more complicated for the strong interaction. Quarks can emit or absorb gluons. Gluons can emit or absorb gluons. Gluons can split into virtual quark-antiquark pairs. These virtual particles can undergo further interactions. This leads to wild tangle of gluons, quarks and their antiparticles flickering in and out of existence as they are emitted and reabsorbed.

The left panel of Fig. 2.1 shows a proton and the middle panel a neutron. The coloured circles represent the quarks (the colours are arbitrary, all that matters is all three are present so overall nucleons are color neutral) and the curly lines represent gluons carrying the strong interaction. For clarity, these two diagrams only show direct interactions between the quarks. The right panel includes some of the other possible types of interactions. Now imagine an endless avalanche of these interactions and you can see the challenge of solving QCD problems.[2]

What does all this have to do with emergence? Quarks have fractional charge. Protons have integer charge. Quarks have color and experience the strong interaction directly. Nucleons are color neutral. In these senses, more is different. But that's not all. Unless you are a high energy physicist, none of

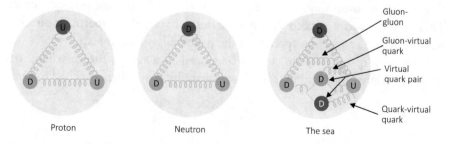

Fig. 2.1 Three quarks for Muster Mark!

---

[2] For a beautiful visual representation, see https://arts.mit.edu/projects/visualizing-the-proton/.

this matters. You can do everything you need to do, including nuclear fission and fusion, while treating protons and neutrons as fundamental particles, as the featureless grey circles in the figure. The approximation involved only breaks down at high energy levels. This is why nucleons were thought to be fundamental right up to the 1960s. Nucleons are nothing more than their components, yet for all practical purposes they are independent of them.

There are two further senses in which the interactions of quarks lead to emergence. Atomic nuclei are bound together by the nuclear force which overcomes the electromagnetic repulsion between positive protons. Yet this force is just a residual of the strong interaction between quarks, orders of magnitude weaker than the strong interaction itself. The nuclear force is emergent.

Then there's the question of mass. If you're up to speed on physics and are asked where mass comes from, you would probably answer that it's to do with the Higgs field. But you'd be mostly wrong. The mass of a proton is around 140 times the mass that the Higgs field produces for its three quarks. The rest of the mass comes from the energy of the cloud of virtual particles shown in the right panel of Fig. 2.1. This is known as emergent hadron mass. Pause for a minute to think about this. Around 98% of the mass of the visible universe is emergent in this sense.[3]

Right down at the heart of matter, we've already got three emergent phenomena. Nucleons emerge from their component quarks. Their masses emerge from the interaction between these quarks. And the nuclear force which holds nuclei together emerges from the strong interaction.[4]

2.2   The Classical World

Quantum physics describes systems by a wave function. One implication of this is that quantum systems are simultaneously in all their possible states. This is known as a superposition of states, or simply a superposition. The wave function can be interpreted as the probability of each state.

Imagine a quantum coin. It can be placed in a superposition where it is simultaneously heads and tails, with a probability of one half attached to each.

---

[3] Binosi (2022).

[4] In fact, all the properties of nucleons are emergent. The figure shows them as shaded grey circles, but their measurable radius is a consequence of the nature of the strong interaction. Their spin also emerges in some complex way from the spin of their component quarks and gluons.

The better-known example is Schrödinger's cat which is in a superposed state consisting of awake and asleep.[5]

We never observe such superpositions. Instead, we experience a world where objects are in one state at a time. When we toss a coin, we see heads or tails. Cats are either awake or asleep. How can we reconcile this classical world with the quantum world that underlies it?

One answer is quantum decoherence. Let me illustrate it by continuing with the example of a coin. The left panel of Fig. 2.2 shows a classical coin, either heads or tails. In the middle is a quantum coin, prepared in a superposition between heads and tails. The quantum coin is shown inside a perfectly empty box. To preserve the superposition, or more precisely to preserve its coherence, the coin must be kept isolated from its environment. Coherent superpositions are extremely fragile.

In the world, quantum systems are not isolated but in environments full of particles and radiation. These scatter off the quantum coin, become entangled with it and the coherence of the superposition leaks away into the environment. This is shown in the right panel of the figure.

For macroscopic objects, decoherence happens extremely quickly.[6] Due to the effects of sunlight alone, a speck of dust would decohere in 10^{-12}s and a bowling ball in 10^{-20}s. Even in the ultra-pure vacuum of deep space, the photons of the cosmic microwave background would cause decoherence of dust in 10^{-4}s and the ball in 10^{-15}s. This is why we never observe superpositions. It is also one of the reasons why quantum computing, which depends on such superpositions being maintained, is a challenge.

A quantum system open to its environment behaves dramatically differently from an isolated quantum system. More is different. The classical world

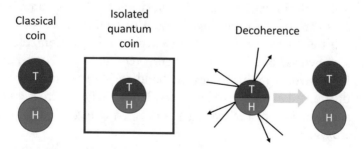

Fig. 2.2 Decoherence

---

[5] I borrow this gentle formulation from Rovelli (2021).
[6] See Appendix A.2 for details of the calculation.

we experience is emergent and decoherence explains how it dynamically emerges from the quantum world.

2.3   Atoms and Molecules

The properties of nucleons depend on their environment. Isolated neutrons are unstable. Due to the weak interaction, they decay with a half-life of around 15 min into a proton, an electron and an antineutrino. Isolated protons, on the other hand, are either stable or have extremely long half lives. Inside light nuclei, protons and neutrons are both stable. Inside heavier nuclei, protons can decay by positron emission, again due to the weak interaction. This is emergence. You cannot understand the properties of a nucleon without understanding its environment.

Now let's turn to atoms. The simplest atom is hydrogen, consisting of a proton and an electron. By themselves, these particles just get on and do their own thing. Combined, they give the atom a whole range of interesting new properties. Most notably, the electron becomes confined in what are known as orbitals. Some of these are illustrated in Fig. 2.3.

When photons scatter off the atom, transitions between these orbitals give characteristic spectral lines. All these properties are dramatically different from those of an isolated proton or electron. Understanding the behaviour of the particles without taking into account the atomic environment is impossible.

Atoms combine to form molecules and the molecules have properties different from their components. Let's take water as an example. The water molecule is composed of two hydrogen atoms bound to one oxygen atom. This is shown in the inset of Fig. 2.4. The nucleus of an oxygen atom contains eight protons compared to the single proton of hydrogen. This means that the molecule's eight electrons, six from oxygen and one from each of the hydrogens, shown as black dots on the figure, tend to be closer to the oxygen nucleus. This, when combined with the bond angle of around 105°, means charge is distributed asymmetrically across the molecule. There is a net positive charge on the side of the hydrogen atoms, a net negative charge on the side of the oxygen atom.

This allows water molecules to form bonds with each other, the negative charge on the oxygen atom in one molecule being attracted to the positive charge on the hydrogen atom in another molecule. This is shown in the main part of the figure. The oxygen atoms are in red, the hydrogen atoms are in grey. The dotted lines representing the electronic attraction between them.

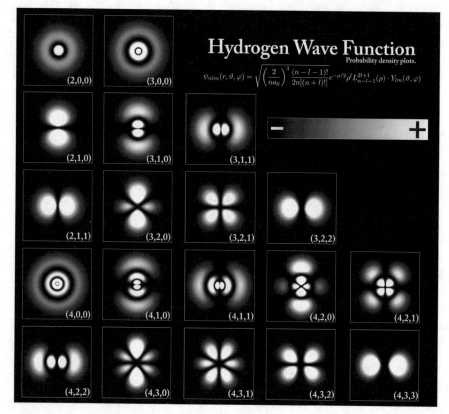

Fig. 2.3 Hydrogen orbitals[7]

Such bonds are known as hydrogen bonds and are responsible for many of the unique properties of water. Hydrogen bonds and the properties of water are emergent.

And so on to the rest of chemistry. Here we've seen three levels of emergence, three levels at which more is different: in the nucleus, in the atom and in molecules.

2.4 Chemical Oscillators

Mix most chemicals and, if they react at all, they will rapidly reach equilibrium. In 1951, Russian chemist Boris Belousov showed that if a particular set of chemicals are mixed in a beaker, the liquid starts off colourless, changes

---

[7] Source: https://commons.wikimedia.org/wiki/File:Hydrogen_Density_Plots.png. License: Public domain.

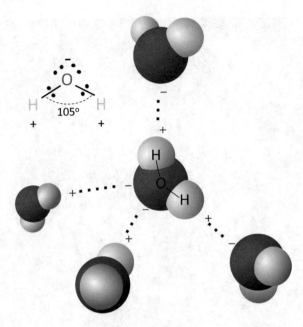

Fig. 2.4 Water[8]

to yellow, turns colourless again, changes back to yellow and so on. Instead of quickly reaching equilibrium like all chemical reactions then known, this cycle can be repeated for up to an hour. Belousov had discovered a chemical oscillator.

He tried to publish his results twice. In response to his first article, in 1951, the editor advised him that his "'supposedly discovered discovery' was quite impossible".[9] After much further work, he submitted another article in 1957, but faced similar scepticism and resolved not to publish. Another Russian chemist, Anatol Zhabotinsky, published a description of the reaction in 1964 and the reaction is known as the Belousov-Zhabotinsky (BZ) reaction.

One variant of the BZ reaction involved mixing chemicals in a petri dish. They initially have a uniform pink colour. After a while, blue spots form and start growing. Then within the blue spots, pink spots form and start growing. Within these pink spots, new blue ones start growing and so on forming a complex and constantly changing pattern. Figure 2.5 shows an example.[10]

Here is Ilya Prigogine's (Nobel Prize for Chemistry, 1977) description:

[8] Source: https://commons.wikimedia.org/wiki/File:3D_model_hydrogen_bonds_in_water.svg. License: Public domain.

[9] Winfree (1984).

[10] A video can be found at www.TheMaterialWorld.net.

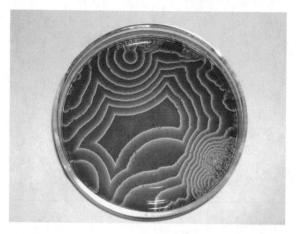

Fig. 2.5 The BZ reaction[11]

Such a degree of order stemming from the activity of billions of molecules seems incredible, and indeed, if [such reactions] had not been observed, no one would believe that such a process is possible. To change color all at once, molecules must have a way to 'communicate.' The system has to act as a whole.[12]

Once more, this is emergence. If the system acts as a whole, you can't understand the behaviour of the individual atoms without understanding the behaviour of the system.

2.5 Symmetry Breaking

Figure 2.6 shows a ball at the top of a hill, with identical valleys on either side. The setup is perfectly symmetric, the physics of both sides are exactly the same. However, when the ball rolls one way or the other, this symmetry is lost.

That is the basic idea of symmetry breaking. Some event, perhaps a random nudge from the molecular storm, changes a symmetric initial state into an asymmetric final state. In this highly stylised example, that's the end of the story. In general, symmetry breaking will have wider effects.

---

[11] Source: https://www.flickr.com/photos/nonlin/. By Stephen W. Morris. License: Creative Commons Attribution 2.0 Generic.
[12] Prigogine and Stengers (1984), p. 148.

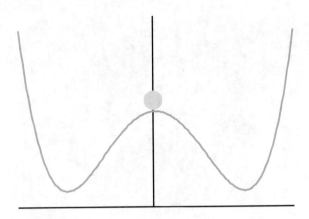

Fig. 2.6 Symmetry breaking

Phase transitions are an example of symmetry breaking. Imagine steam cooling to form liquid water. Steam is highly symmetric in the sense that each molecule is independent of all the others. This symmetry is broken when steam condenses into liquid water and hydrogen bonds (Fig. 2.4) form between the molecules.

In general, high temperature phases have more symmetries than low-temperature phases. As a system is cooled, it may go through multiple phase transitions breaking symmetries at each of them. What's more, phase transitions appear discontinuous, water molecules at a fraction of a degree below boiling point behave differently than water molecules at a fraction of a degree above. There are many different types of phase transitions: between different states of matter (solid, liquid and gas); in magnetic substances between different kinds of magnetism and in solids between different crystal structures. Models of cosmology describe the period after the big bang as a series of phase transitions, progressively breaking the symmetries of the very early universe.

Phase transitions only happen if there are a large number of particles. If we think of individual molecules, all that happens as temperature falls is that they move more slowly. To see boiling and freezing, we need to look at the system as a whole. More is different. Phase transitions are emergent phenomena.

2.6   Quasiparticles

Condensed matter physics studies solids and liquids. The dynamics of such systems are extremely complicated, with huge numbers of nuclei and electrons interacting with each other. This complexity would seem to make modelling them difficult. However, the collective behaviour of such systems can sometimes be treated as if it were a particle. Instead of having to keep track of countless trillions of nuclei and electrons, systems can be described by a few such quasiparticles.

The quasiparticle approach was first used in the 1930s. Since then, it has been applied to many different phenomena. Wikipedia lists around 30. Electron holes, resulting from the absence of an electron, behave like positively charged particles. Magnons are collective excitations of electron spins in a magnetic material. Phonons are vibrational modes of a crystal. Let's look at them in more detail.

Atoms in a crystal lattice are held together by interatomic forces. We can think these as like springs, as illustrated in the left panel of Fig. 2.7. If you give the crystal a bang on its left side, the lattice will start vibrating and the net result will be to transfer the energy delivered by the bang from the left to the right. The atoms in the lattice stay in approximately the same place during this process.

The vibrations can be divided up into *normal modes*, just like for a vibrating string. This is illustrated on the right side of the figure. Any vibrational pattern of the lattice can be expressed as a combination of normal modes. And each normal mode can be treated as a particle, a phonon.

If we give the lattice a bang on the left such that all the atoms vibrate with the same frequency, we can describe this as a single phonon moving through the lattice, transferring energy from left to right. If the bang involves multiple frequencies, the description will involve multiple phonons. Although all the atoms in the lattice are vibrating in complex ways, their behaviour can be captured by a small number of phonons.

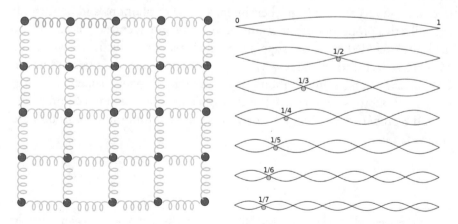

Fig. 2.7 Phonons

The particles of light, photons, are quantised in that their energy can only change in discrete steps. The same is true of phonons. You can think of them as quanta of sound. If we can bang the lattice in such a way as to produce a single phonon, a sequence of such bangs would produce a coherent beam of phonons, the sound equivalent of a laser. Heat, the random vibrations of the lattice, produces incoherent phonons of many frequencies moving in many directions.

This might seem like magic. Rather than having to solve the physics of a hugely complex system, we have managed to describe it in dramatically simpler terms.[13] Is it a trick? Phonons seem as real as other particles. You can scatter them off each other. If we bang the lattice at the top as well as on the left, this would produce two streams of phonons, one going from left to right and the other from top to bottom. Their interaction would be governed by the law of conservation of phonon momentum. Phonons can also scatter off electrons or neutrons and can even be shown to be fermions and so subject to quantum indistinguishability.

Far from being an abstraction of theoretical physicists, the discipline of phononics underlies many technological advances. Designing materials to have particular properties means understanding the behaviour of their phonons. This is from a 2014 review:

> Understanding and controlling the phononic properties of materials provides opportunities to thermally insulate buildings, reduce environmental noise, transform waste heat into electricity and develop earthquake protection.[14]

Another example is superconductivity. Materials become superconducting when their electrons couple up to form Cooper pairs. This happens despite electronic repulsion thanks to their interaction with phonons and hence with the collective properties of the atomic lattice. A consequence is that magnetic fields are excluded from the superconducting material. This is known as the Meissner effect and explains why magnets levitate above superconductors. One of the stranger implications of this is that, inside a superconductor, the virtual photons which carry the electromagnetic interaction acquire mass. Even the properties of photons cannot be separated from the systems they are embedded in.

---

[13] For the purposes of exposition, I've described phonons in classical terms. To properly model the vibrations of a crystal lattice, you need quantum physics.

[14] Maldovan (2013).

And these are just two examples from dozens. It's not hard to see why condensed matter physics is often viewed as the most fertile ground for concepts of emergence.

2.7   The Quantum Hall Effect

The Hall effect was discovered at the end of the nineteenth century. Pass a current through a thin metal sheet then apply a magnetic field perpendicular to the sheet. The electrons carrying the current experience the magnetic field as a force, called the Lorenz force, pushing them perpendicular to the direction of the current. So they tend to accumulate on one side of the sheet creating a potential difference, known as the Hall voltage. This is illustrated in the left panel of Fig. 2.8.

Divide this voltage by the current and you get the Hall resistance. The stronger the magnetic field, the larger the Hall resistance. Plot the field against the resistance and you get a straight line as in the right panel of the figure.

In 1980, Klaus von Klitzing (Nobel Prize for Physics, 1985) discovered the quantum Hall effect. Instead of a thin metal sheet, this involves electrons confined to a two-dimensional material, such as the interface between two semiconductors or a sheet of graphene. It also requires low temperatures (4K) and high magnetic fields (10T).[15] Under these conditions, the relation between the magnetic field and the Hall resistance is no longer linear but has

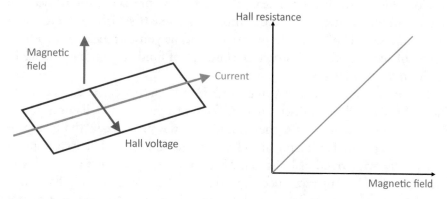

Fig. 2.8   Hall effect

---

[15] A note on units. Temperatures are measured in Kelvin (K). A Kelvin is the same as a degree centigrade, but the scale starts at absolute zero. So 0 K = −273 °C. Magnetic fields are measured in Tesla (T). For a comparison, the strength of the earth's magnetic field is around 10^{-5} T; the magnet in a loudspeaker is around 1 T.

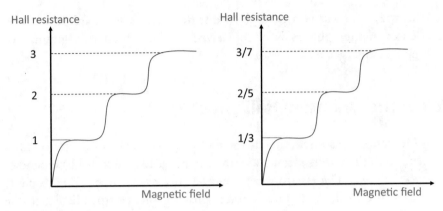

Fig. 2.9 Quantum Hall effect

regular plateaux. This is shown in the left panel of Fig. 2.9. What is extraordinary is that the levels of these plateaux are integer multiples of a combination of fundamental constants.[16]

More extraordinary still is that the quantisation is extremely precise, down to the limits of experimental accuracy and that this precision is independent of most of the properties of the material. Neither size, shape nor impurities in the sample matter. Indeed, it requires some impurities as the effect disappears in pure samples. This is the sense in which the quantum Hall effect is described as emergent, it seems to float free of almost all of the properties of its substrate.

Does this sound abstract? The measurement is so precise that it has become a standard for electrical resistance. Also, given some basic lab equipment and an impure sample of your 2D material, it means you can measure to a high degree of accuracy the fine-structure constant, of fundamental importance in quantum electrodynamics.

Two years later, Horst Stormer and Dan Tsui (who shared the 1998 Nobel Prize for Physics) discovered the fractional quantum Hall effect. Using specially prepared materials, they found that the plateaux in the quantum Hall effect appeared at fractional and not integer multiples. These correspond to particles with fractional charges. When the first plateau was found at a multiple of 1/3, Tsui exclaimed "Quarks!".[17] What had actually happened was the electrons had undergone a phase transition into a new state of matter which can be described by quasiparticles with fractional charges.

---

[16] The plateaux occur at multiples of h/e^2 where h is the Planck constant and e the charge on the electron.

[17] Stormer (1999).

The fractional effect is just as robust to the properties of the underlying material. In 1982, David Thouless gave an explanation of this in terms of topology showing that the effect depends only on the general topological properties of the material. Think of a doughnut. You can distort it in lots of ways but it still stays a torus. It takes much more effort to change the topology, by tearing it or punching new holes. Thouless showed that, whereas ordinary phase transitions are characterised by symmetry breaking, the fractional quantum Hall phase is topologically distinct. He shared the 2016 Nobel Prize for Physics with Duncan Haldane and Michael Kosterlitz who also pioneered the application of topology to physical systems. Since then, the study of topological effects has become an exciting new area of physics. A whole range of systems can be classified by their topologies, the properties of the system as a whole, independent of the details of their structure.

Robert Laughlin (the third laureate of the 1998 Nobel Prize for Physics). is unequivocal about the importance of this:

> I have come to understand the von Klitzing discovery to be a watershed event, a defining moment in which physical science stepped firmly out of the age of reductionism into the age of emergence.[18]

2.8   Bénard Convection

Take a stylised version of a kettle. An insulated cylinder contains a liquid. The base of the cylinder is held at a temperature higher than the top. How does the liquid behave? Its molecules pick up kinetic energy from the hot base and through collisions transfer this kinetic energy to other molecules. When a molecule collides with the cool top, it loses kinetic energy. The result is a net transfer of energy from bottom to top, a flow of heat. This is conduction. If you increase the temperature gradient, heat flows faster.

But when the gradient reaches a critical level something unexpected happens. A highly structured pattern of hexagonal cells appears. Viewed from above, the liquid looks like a beehive. This is shown in the left panel of Fig. 2.10. The image is adapted from the 1901 article in which Henri Bénard first described the phenomenon.

The middle panel of the figure shows the flow from the side. Within these Bénard cells, heat transfer is by convection, smooth flows of fluid rather than the random molecular motion of conduction. The fluid has spontaneously

---

[18] Laughlin (2005), p. 76.

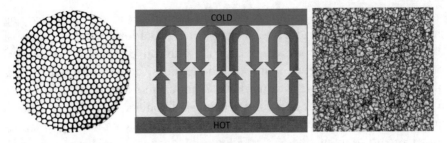

Fig. 2.10 Bénard convection[19]

adopted a pattern that transfers energy faster than would conduction alone. Somehow, the individual molecules have acted collectively to produce this large-scale pattern. Such convection cells appear wherever there are fluids and temperature gradients. They are behind the granular appearance of the surface of the sun, shown in the right panel of the figure.

You can see this as a phase transition from conduction to convection which breaks the symmetry of the liquid. When there is convection, it is impossible to understand the behaviour of a molecule in one of the convection cells without understanding the behaviour of the cell as a whole. And it impossible to understand a single convection cell without understand its place in the overall pattern.

2.9   Self-organisation

The last four examples are often described as self-organising systems. All are characterised by the emergence of order from a decentralized interaction between their component atoms or molecules. Hermann Haken gives a simple analogy to explain self-organisation:

> Consider, for example, a group of workers. We then speak of organization or, more exactly, of organized behavior if each worker acts in a well-defined way on given external orders... We would call the same process as being self-organized if there are no external orders given but the workers work together by some kind of mutual understanding, each one doing his job to produce the product.[20]

---

[19] Left panel: Bénard (1901). Right panel: Source: https://nso.edu/gallery/. Taken by the Daniel K. Inouye Solar Telescope. License: Creative Commons Attribution 4.0 International (CC BY 4.0).

[20] Haken (1983), p. 191.

Fig. 2.11 Self organisation[21]

Whether it's this "mutual understanding", or the "communication" that Prigogine saw among the molecules of chemical oscillators, it's hard to avoid the conclusion that something extraordinary is going on.

Figure 2.11 gives three more examples. The pristine geometry of snowflakes, the Fibonacci spiral of the sunflower, the endlessly shifting dynamics of a murmuration of starlings. All these emerge spontaneously from the interaction of their parts. And all exhibit properties different from those of their parts.

2.10 Ordinary Objects

Look around you. Everything you see is made up of atoms. Yet this is irrelevant for most purposes. It is irrelevant to the extent that the atomic view of the world only became accepted among physicists in the first decades of the twentieth century. The nature of objects is robust to changes in their atomic make up. Swap round a few atoms and the object will be unchanged. It's a philosophers' game to argue exactly how many you can change before the object changes identity.

Objects have properties that their constituent atoms don't: they may be round, red, hard. They obey regularities different from those in the atomic realm. Throw a cricket ball and it will describe a parabola (we'll look at the problem of how to catch it in Chap. 4). Throw a cricket ball and it may smash a window. Such regularities are mostly independent of what the ball is made

---

[21] Left panel: Warren (1863), p. 39. Middle panel: Source: https://pxhere.com/en/photo/659385. License: CC0, Public domain. Right panel: Source: https://commons.wikimedia.org/wiki/File:Starling_flock_with_nearby_predator.jpg. License: Creative Commons Attribution 4.0 International (CC BY 4.0).

of. Any rigid sphere of sufficient density will travel in a parabola and smash windows. Whether leather, wood, metal or stone, whatever the ball is made of, to understand the behaviour of one of its atoms you need to understand its ball-like nature.

Examples of such high level regularities are everywhere in our daily lives. Disciplines such as geology, psychology or economics describe systems of mindboggling complexity but manage to identify useful regularities. For objects and the laws which govern them, more seems decidedly different.

2.11   Game of Life

No example is more beloved of proponents of emergence than John Conway's Game of Life. It consists of a grid of square cells, so each cell has 8 neighbours. A cell can either be alive, shown as black on the grid, or dead, shown as white. At each tick of a clock, the following rules are applied to squares on the grid:

1. A live cell with fewer than 2 live neighbours dies ("of loneliness")
2. A live cell with 2 or 3 live neighbours lives on
3. A live cell with more than 3 live neighbours dies ("of overcrowding")
4. A dead cell with 3 live neighbours becomes alive ("reproduction).

All you need to do is specify an initial state and set the program running. The result is a whole zoo of complex behaviour. Figure 2.12 shows a pattern known as a glider. After 4 periods it moves one square diagonally downwards and to the right and will continue doing so until it bumps into other live cells or the edge of the grid. There are more complex patterns known as glider guns which produce a constant stream of gliders. If you haven't already done so, I do suggest you have a play with one of the versions available online.[22]

The Game of Life provides an ideal environment in which questions of emergence can be explored. In the game, a set of simple, well-defined

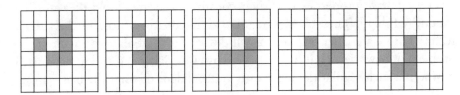

Fig. 2.12   A glider

---

[22] A link is at www.TheMaterialWorld.net.

rules lead to complex behaviour. The parallel with other systems is seductive. Perhaps such simple rules are behind all self-organisation. Perhaps the (relatively simple) laws of physics lead to the complexity of the universe.

Are there limits to the complexity of systems that these simple laws can produce? If so, they have yet to be found. In 2010, New Scientist magazine[23] reported the creation of a self-replicating creature within the Game of Life.

2.12 Evolution

Let's turn from the Game of Life to life itself. The proteins in your body are combinations of 20 different amino acids. A short protein might contain 200 different amino acids and so 20^{200} such proteins are possible. This is a number so large it defeats any comparison. Stuart Kauffman writes:

> ...if the 10 to the 80th particles in the universe were doing nothing since the Big Bang except making proteins in parallel at every tick of the Planck time clock, it would take 10 to the 39 power times the 13.7-billion-year actual history of the universe to make all possible proteins of the length 200 amino acids, just once.[24]

This implies that there is path dependence. Out of a myriad possible paths leading to a myriad different outcomes, just one has been arbitrarily selected. If we try to build a model of protein evolution, every set of initial conditions will lead us to a different outcome. We'll never be able to explain why life on earth uses particular proteins because it is a result of arbitrary and possible infinitesimal events billions of years ago.

Kauffman argues that this applies to every aspect of evolution. The biological world we see is the result of a long sequence of symmetry breaking going right back to the start of time:

> We can write no laws of motion ... for the emergence of the eukaryotic cell, sex, multicelled organisms, the Cambrian explosion with its specific marvels of the explosion of diversity of early flora and fauna, promissory of us, fish, amphibians, reptiles, mammals, and primates, let alone the specific proteins that have emerged. We live in an unprestatable, literally unimaginable, myriad of emergent becoming. Because we can write no laws for the specific emergence we call life, we are based on physics, but beyond physics.[25]

---

[23] Aron (2010).
[24] Kauffman (2019), p. 3.
[25] Kauffman (2019), p. 127.

This is emergence on a grand scale. There are causal chains that are simply too complex for us to untangle. Living systems are collections of atoms but are beyond physics. More is resoundingly different.

2.13 Living Cells

Inanimate objects are already fascinating. But a simple bacterium is at a different level. Bacteria show a remarkably diverse range of behaviour including sensing their environment, learning and prediction, the generation of circadian rhythms, signalling to their fellows and forming collectives. Let's focus on one such bacterium, the well-known Escherichia coli or E. coli and look at the way it can search for food.

Chemotaxis is the ability to follow chemical gradients. Figure 2.13 gives a stylised illustration. Somewhere off the bottom of the image is a chemical that will be useful to the bacterium, call it food. This diffuses into the environment resulting in a concentration increasing from the top to the bottom of the image.

The bacterium has two ways of moving. If its flagella turn counterclockwise, it proceeds in a straight line. If they turn clockwise, it stays in one place but changes its orientation. The first is called swimming, the second tumbling.

The figure shows the bacterium swimming from bottom left to top centre. It senses the decreasing chemical gradient. To do this requires a way of detecting the chemical, a memory of what the concentration was in the past

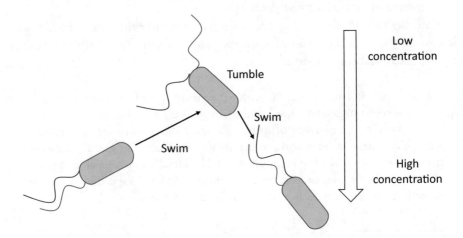

Fig. 2.13 Swim and tumble

and a way of comparing this with the current concentration. At some point, this causes the direction of the flagella to reverse and the bacterium starts to tumble. The tumble ends at a random orientation and the bacterium swims on towards the bottom right. The likelihood of a tumble is much higher if the concentration is decreasing than if it is increasing. This will mean that, on average, the bacterium approaches the food.

The bacterium is emergent in the same sense as are ordinary objects. But there is something more. The bacterium is controlling its behaviour to achieve a goal. This combination of top-down control and goal directedness is a new level of emergence.

2.14 Turning the Page

Turn back a few pages and remind yourself of what I wrote about QCD.

Did you do it? If so, you acted with purpose. And enormous numbers of protons, neutrons and electrons just did what they were told. We have a thought, and this thought changes the world. Such purposiveness is a unique property of minds; fundamental particles do not have it. More is different.

What's more, whether you turned the page or not depended on your disposition towards bossy authors. The motion of all those fundamental particles depended on your mental state. You can't understand the physical without the mental.

Philosophers use the term *intentionality* for this ability of minds to refer to something, to be about something. And many philosophers think this must put minds beyond physics. Here is Jerry Fodor:

> I suppose that sooner or later the physicists will complete the catalogue they've been compiling of the ultimate and irreducible properties of things. When they do, the likes of spin, charm, and charge will perhaps appear upon their list. But aboutness surely won't; intentionality simply doesn't go that deep...
>
> ...there is no place for intentional categories in a physicalistic view of the world; ... the intentional can't be naturalized. It is time that we should face this issue. What is it, then, for a physical system to have intentional states?[26]

Note that this point holds whatever you think about the slippery notions of free will and consciousness. There is a direct chain of causation from the content of mental states to the displacement of fundamental particles.

---

[26] Fodor (1998), p. 97.

2.15 Churchill's Nose

David Deutsch asks us to consider the statue of Sir Winston Churchill that stands outside the Houses of Parliament in London. At the tip of the statue's nose is an atom of copper. Why is that atom there? Here is Deutsch's explanation:

> …It is because Churchill served as prime minister in the House of Commons nearby; and because his ideas and leadership contributed to the Allied victory in the Second World War; and because it is customary to honour such people by putting up statues of them; and because bronze, a traditional material for such statues, contains copper, and so on.[27]

This is an explanation in terms of concepts which have no place in the world of physics: leadership, war and tradition. To explain the position of an atom, a basic physical fact, we need a whole constellation of emergent concepts. Deutsch accepts that, given enough computing power and scientific knowledge, we might be able to calculate the probability of a copper atom being there. But he asserts that such a prediction would still not explain anything. Physics is not enough. To understand the world, emergent concepts are essential.

2.16 Common Themes

Throughout this chapter, I've sketched ideas which will come up in the rest of the book. The difficulty of prediction. The way in which macroscopic phenomena seem independent of their microscopic constituents. And the constant theme that you can't understand individual parts without considering their environment. Each of these is a type of weak emergence and are the subjects of Chaps. 7–9. Before that, Chap. 6 will investigate "more is different".

Then there is a question which may have crossed your mind as you read. Is current physics enough to explain these examples? Or is new physics needed? This brings us on to strong emergence, which is the subject of Chap. 11.

More generally, we've seen that emergence can be found in systems ranging from atomic nuclei to statues. Here's a question for you: can you think of a system which is not emergent?

---

[27] Deutsch (1998), p. 22.

These examples also show that emergence involves dividing the world into levels. Let's return to the atom of Churchill's nose. Going downwards, you can see the atom as emerging from the combination of its nucleus and electrons, the nucleus emerging from the combination of protons and neutrons and these emerging from their constituent quarks and gluons. Going upwards, the atom is part of a nose, and the nose part of the statue, both ordinary objects, and above that is the whole framework of life, mental causation and the evolutionary process which leads to them. This layered picture of the world is where the next chapter starts.

2.17 Further Reading

For a general overview of emergence, see the entry in the Stanford Encyclopaedia of Philosophy, O'Connor (2021). Two excellent collections of articles are Falkenburg and Morrison (2014) and Gibb (2019).

If you want a brief introduction to any of the examples covered in this chapter, I suggest you start with a standard textbook or the Wikipedia entry. For each topic, here is a paper or book I found particularly useful. Protons and neutrons: Marciano and Pagels (1979); the classical world: Joos (2006); Atoms and molecules: Esposito and Naddeo (2013); Chemical oscillators: Epstein and Showalter (1996); Symmetry breaking: Brading et al. (2023); Quasiparticles: Venema et al. (2016); Bénard convection: Manneville (2006); Self-organisation: Camazine (2003). Ordinary objects: Thomasson (2010); Game of life: Conway (2004); Evolution: Kauffman (2019); Biological systems: Berg (2004); Brains and minds: Dennett (2017); Churchill's nose: Deutsch (1998).

More suggestions for reading can be found at www.TheMaterialWorld.net.

References

Anderson PW (1972) More Is Different: Broken symmetry and the nature of the hierarchical structure of science. Science 177:393–396. https://doi.org/10.1126/science.177.4047.393

Aron J (2010) First replicating creature spawned in life simulator. New Scientist

Bénard H (1901) Les tourbillons cellulaires dans une nappe liquide. - Méthodes optiques d'observation et d'enregistrement. J Phys Theor Appl 10:254–266. https://doi.org/10.1051/jphystap:0190100100025400

Berg HC (2004) E. coli in motion. Springer, New York

Binosi D (2022) Emergent Hadron Mass in Strong Dynamics. Few-Body Syst 63:42. https://doi.org/10.1007/s00601-022-01740-6

Brading K, Castellani E, Teh N (2023) Symmetry and Symmetry Breaking. In: Zalta EN, Nodelman U (eds) The Stanford Encyclopedia of Philosophy, Fall 2023. Metaphysics Research Lab, Stanford University. https://plato.stanford.edu/entries/symmetry-breaking/

Camazine S (ed) (2003) Self-organization in biological systems. Princeton University Press, Princeton, NJ

Conway JH (2004) What is Life? In: Winning ways for your mathematical plays. 4. Peters, Wellesley, Mass

Dennett DC (2017) From bacteria to Bach and back: the evolution of minds. W.W. Norton & Company, New York London

Deutsch D (1998) The fabric of reality: the science of parallel universes-- and its implications. Penguin Books, New York

Epstein IR, Showalter K (1996) Nonlinear Chemical Dynamics: Oscillations, Patterns, and Chaos. J Phys Chem 100:13132–13147. https://doi.org/10.1021/jp953547m

Esposito S, Naddeo A (2013) The genesis of the quantum theory of the chemical bond. https://doi.org/10.48550/ARXIV.1309.4647

Falkenburg B, Morrison M (2014) Why more is different. Springer, New York

Fodor JA (1998) Psychosemantics: the problem of meaning in the philosophy of mind. MIT Pr, Cambridge, MA

Gibb SC (ed) (2019) The Routledge handbook of emergence. Routledge, New York

Haken H (1983) Synergetics: an introduction: nonequilibrium phase transitions and self-organization in physics, chemistry, and biology. Springer, Berlin ; New York

Joos E (2006) The Emergence of Classicality from Quantum Theory. In: Clayton P, Davies PCW (eds) The re-emergence of emergence: the emergentist hypothesis from science to religion. Oxford University Press, Oxford; New York

Kauffman SA (2019) A world beyond physics: the emergence and evolution of life. Oxford University Press, New York, NY

Laughlin RB (2005) A different universe: reinventing physics from the bottom down. Basic Books, New York

Maldovan M (2013) Sound and heat revolutions in phononics. Nature 503:209–217. https://doi.org/10.1038/nature12608

Manneville P (2006) Rayleigh-Bénard Convection: Thirty Years of Experimental, Theoretical, and Modeling Work. In: Mutabazi I, Wesfreid JE, Guyon E (eds) Dynamics of Spatio-Temporal Cellular Structures. Springer New York, New York, NY, pp 41–65

Marciano W, Pagels H (1979) Quantum chromodynamics. Nature 279:479–483. https://doi.org/10.1038/279479a0

McLaughlin BP (2008) The rise and fall of British emergentism. In: Bedau MA, Humphreys P (eds) Emergence: contemporary readings in philosophy and science. MIT Press, Cambridge, Mass

Noordhof P (2019) Dependence. In: Gibb SC (ed) The Routledge handbook of emergence, 1 [edition]. Routledge, New York

O'Connor T (2021) Emergent properties. The Stanford Encyclopedia of Philosophy. https://plato.stanford.edu/entries/properties-emergent/

Penev ES, Marzari N, Yakobson BI (2021) Theoretical Prediction of Two-Dimensional Materials, Behavior, and Properties. ACS Nano 15:5959–5976. https://doi.org/10.1021/acsnano.0c10504

Prigogine I, Stengers I (1984) Order out of chaos: man's new dialogue with nature. Bantam Books, Toronto; New York, N.Y

Rovelli C (2021) Helgoland: making sense of the quantum revolution. Riverhead Books, New York

Seager W (2018) Idealism, Panpsychism, and Emergentism. In: Gennaro RJ (ed) The Routledge handbook of consciousness, 1 [edition]. Routledge, New York

Stormer HL (1999) Nobel Lecture: The fractional quantum Hall effect. Rev Mod Phys 71:875–889. https://doi.org/10.1103/RevModPhys.71.875

Thomasson AL (2010) Ordinary objects. Oxford University Press, Oxford New York

Venema L, Verberck B, Georgescu I, et al (2016) The quasiparticle zoo. Nature Phys 12:1085–1089. https://doi.org/10.1038/nphys3977

Warren IP (1863) Snowflakes: a chapter from the book of nature. American Tract Society, Boston

Winfree AT (1984) The prehistory of the Belousov-Zhabotinsky oscillator. J Chem Educ 61:661. https://doi.org/10.1021/ed061p661

Part I

Frameworks

3

Dualism, Physicalism and Emergence

Summary This chapter provides a philosophical framework for the rest of the book. Emergence is closely linked to the idea that the world is divided into levels, with higher levels emerging from lower ones. Once the concept of levels is clarified, the chapter turns to Jaegwon Kim's argument that emergence is either an empty concept or must contradict physics. The argument can be stated as six propositions of which only a maximum of five can be true. The choice of which proposition to reject corresponds to a way of seeing the world: dualism, three types of physicalism, weak emergence and strong emergence.

The three chapters of Part I introduce ideas which will be used throughout the remainder of the book. This chapter provides a philosophical framework, the next turns to human cognitive evolution and the nature of the scientific project then Chap. 5 addresses issues of computation and simulation. The aim is to make the book largely self-contained but it means that many readers will come across material they are familiar with. In this case, recall that the French author Daniel Pennac claimed that among the inalienable rights of the reader is the right to skip pages.[1] If you'd rather get straight to the discussion of emergence, you can jump to Chap. 6 and refer back as necessary.

---

[1] Pennac (1995).

L. Graham, *Physics Fixes All the Facts*, The Frontiers Collection,
https://doi.org/10.1007/978-3-031-69288-8_3

In the previous chapter I concentrated on examples of emergence and gave no more than a minimum definition. The aim of this chapter is to clarify emergence along with other concepts, such as reduction, eliminativism and physicalism. I'm going to do this in a philosophical framework largely borrowed from a book by philosopher Jessica Wilson.[2] It is based on an argument due to Jaegwon Kim showing that emergent phenomena, if they exist, must contradict physics.

Emergence is a notoriously poorly defined concept. Appendix A.1 contains a (non-exhaustive) list of 75 definitions. The earliest is in this passage from Aristotle:

> For of all things that have several parts and where the totality of them is not like a heap, but **the whole is something beyond the parts**, there is some cause of it, since even among bodies, in some cases contact is the cause of their being one, in others stickiness, or some other attribute of this sort.[3]

From then, there is a gap in the list until John Stuart Mill in the mid-nineteenth century, although one source cites examples throughout enlightenment philosophy. The first decades of the twentieth century saw the flourishing of British Emergentism then things went quiet until the 1970s when work by philosophers and scientists prompted a revival of the concept. All but seven of the definitions are from the past fifty years.

However defined, emergence involves relations between different levels. The examples of the previous chapter are all physical systems, made up of fundamental particles. At higher levels they display emergent properties distinct from those of fundamental particles. Let's start by clarifying this.

3.1   A Layered World

The world we perceive is made up of levels. Wholes are made up of parts. Those parts are made up of smaller parts. We can put this a bit more formally by saying that entities at higher levels are composed of entities at lower levels, but not vice versa. Now for some questions.

What defines a level? There are various ways to answer this, but one that is common in discussions of emergence it to associate a level with a *special science* describing it. The term special science is slightly odd, but it is widely

---

2 Wilson (2021).
3 Aristotle and Reeve (2016), p. 141, emphasis added.

used in the philosophical literature to mean any science that is not fundamental physics. A special science consists of a taxonomy which classifies the entities that form its subject matter, a set of features which these entities possess and a set of laws governing their interaction. For example, the special science of chemistry describes molecules, radicals and ions with features such as valency or electronegativity and laws such as Faraday's Law or Le Chatelier's Principle.

If we think of levels as being associated with a special science, we can read off the levels from a set of academic disciplines: molecules and chemistry; planets and geology; cells and cellular biology; multicellular organisms and zoology; societies and economics.

What is the relation with emergence? Higher levels emerge from lower levels, so all special science features are emergent. There may be higher level features which are not included in a special science, but in what follows I'm going to treat the terms special science and higher level as interchangeable.

What does the hierarchy of levels look like? It is easy to think of a simple, linear hierarchy but a moment's thought shows this cannot be the case. Cells are not made of rocks, but both are made of molecules. People are not made of plants, but both are made of cells. Instead of a simple hierarchy, we have a branching tree. Branches could also rejoin. If you take cognitive science as describing thought processes, it defines a level containing entities that think. At the moment, its only occupants are various animals. But if we one day build an artificial intelligence with similar thought processes, it will belong to the same level. The two distinct branches of the tree, one organic and one inorganic, which split off above the chemistry level, will have rejoined.

We can draw a hierarchy of levels by thinking of relations between the special sciences. Figure 3.1 is taken from a 2008 paper by Max Tegmark. You can, if you choose, use it to read off relations of emergence. Everything emerges from physics. Then there are many other levels of emergence, for example thermodynamics emerging from statistical mechanics or astrophysics from a combination of thermodynamics, chemistry and nuclear physics.

It's hard not to nitpick. Shouldn't there be direct links from quantum mechanics and statistical mechanics to chemistry and biology? Does putting boxes on the same level imply something about the structure of the world? This is not a criticism of the figure since I am using it outside the context for which it was intended. But it does throw into doubt the idea that sciences separate neatly into levels.

A more general question is whether these levels represent the structure of reality or the structure of human knowledge. In philosophical terms, are they ontological or epistemic? I will return to this later in the chapter.

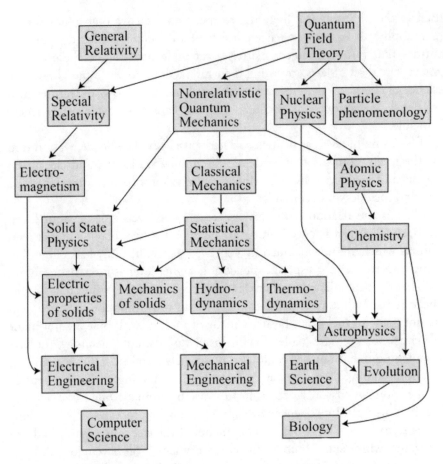

Fig. 3.1 A hierarchy of sciences[4]

3.2 Down at the Bottom

What is the lowest level? That's what the standard model of high energy physics tells us. Everything is composed of particles: five bosons (among them photons) and twelve fermions (among them quarks and electrons). The view that there is some collection of smallest particles can be found in Newton:

> Now the smallest Particles of Matter may cohere by the strongest Attractions, and compose bigger Particles of weaker Virtue; and many of these may cohere and compose bigger Particles whose Virtue is still weaker, and so on for divers

---

[4] Source: Tegmark (2008). https://link.springer.com/article/10.1007/s10701-007-9186-9. Reproduced with permission from Springer Nature.

Successions, until the Progression end in the biggest Particles on which the Operations in Chymistry, and the Colours of natural Bodies depend, and which by cohering compose Bodies of a sensible Magnitude.[5]

Yet this particle-based view of the world may need serious revision. Quantum field theory (QFT) is the most general theory of physics, combining classical field theory, special relativity and quantum mechanics (there is as yet no agreement on how general relativity should be incorporated). However a recent review of the role of particles in QFT starts with the sentence:

> The consensus view among philosophers of physics is that relativistic quantum field theory (QFT) does not describe particles.[6]

This is fascinating in itself, and I will return to it when I discuss quasi-particles in Chap. 15. But for the moment, does this mean that the fields of QFT constitute the lowest level? They are the only candidate and there don't seem to be any obvious objections to having fields at the bottom rather than particles. However, we know that at extremely high energies QFT runs into difficulties and so must be a low energy approximation to a more complete theory. Would this theory be the lowest level? We are unlikely to ever know since there could always be something else lurking just beyond the limits of our experiments.

Is there a lowest level at all? There is no logical reason why there should be. And if there's no lowest level, there must be an infinite number of levels. Leibniz, a contemporary of Newton, held this view and a decent case could be made that it is far more elegant than some arbitrary cut. Karl Popper wrote:

> I do not think that we can ever describe, by our universal laws, an ultimate essence of the world, I do not doubt that we may seek to probe deeper and deeper into the structure of our world or, as we might say, into properties of the world that are more and more essential, or of greater and greater depth.[7]

A further possibility is that there are an infinite number of levels but at some point they become repetitive. While it may be enjoyable to speculate about such issues, there is no empirical evidence one way or another. The hypothesis that there is a lowest level is just as plausible as the hypothesis that

---

[5] Newton (2012), p. 394.
[6] Fraser (2022).
[7] Popper (1995), p. 196.

there is none. From now on, I'm going to use the term going to use the term *microphysical* to refer to the lowest level and assume it is described by QFT which, for all the phenomena covered in this book, can be treated as a Theory of Everything.

3.3   Causation

In what sense are emergent features distinct from their microphysical base? Although there are various ways to address this question, I'm going to follow Wilson's argument that the key property is causation. She frames the discussion in terms of causal powers. A causal power "is simply shorthand for talk of what causal contributions possession of a given feature make".[8] This captures the idea that what entities can do depends on their properties. A stone possesses a causal power, for example smashing a window, because of its properties, velocity, direction, hardness etc. Wilson asserts that this definition has the advantage that is agnostic as to the nature of causation. You should be able to accept it whether you are a sceptic like Hume and think causation is nothing but an empirical regularity or follow Kant and argue that causation is an a priori principle.

If causal power only exists at the lowest level, we have *Physical Causal Closure*, all physical states have purely physical causes. On the other hand, an emergent feature will be distinct from its microphysical base if it has different causal powers. With this clarified, we can turn to the argument that motivates the rest of the chapter.

3.4   Emergence or Physics?

This section presents the *causal exclusion argument* due to Jaegwon Kim.[9] It shows that emergence and physics cannot coexist. Either you accept emergence, in which case you must reject physical causal closure. Or you accept physical causal closure, in which case you must deny that higher levels have causal powers beyond those they inherit from physics.

Take a system with lowest-level properties P (for physics) and higher level properties H. What is the relation between P and H? Let's assume it is an "if

---

[8] Wilson (2021), p. 32.

[9] Kim made the argument in various papers from the 1980s onwards. A useful reference is Kim (2007), Chap. 1.

and only if" so whenever P is observed, H is observed and whenever H is observed, P is observed.

Then let the system changes to a new state given by P* and H*. This is shown in the left panel of Fig. 3.2. Vertical arrows represent the "if and only if" relation between lower and higher levels. Since P is physics, the transition to P* is governed by the laws of physics, indicated by a horizontal arrow, in bold to represent causation. Once physics has produced P* the "if and only if" relation automatically gives H*. This is a physicalist's view of the world with all causation happening at the lowest level.

Now, one definition of reality is having causal power, so if the higher level is real we can say H causes H*. This can happen in two ways. Either H causes H* directly, shown in the middle panel or H causes P* which then automatically instantiates H*, shown in the right panel. This is a special science or emergent view of the world with higher levels having causal power.

Kim's argument is that this leads to a contradiction. In the first case, H* is caused by both P and H. In the second, P* is caused by both P and H. But nothing can have two independent simultaneous causes.[10] So you need to choose. Either you accept H is the cause and the laws of physics are contradicted. Or you accept P as the cause and that there is no causation at higher levels. If reality is having casual power, this means everything except physics is an illusion.

This is a strong result. Fundamental physics has proved astonishingly accurate, so far passing every empirical test. If this leads you to reject anything that contradicts physics, Kim shows you must also reject giving any causal power and so any reality to higher levels. The regularities and laws of the special sciences are no more than window-dressing for physics. Emergence must be an illusion.

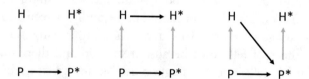

Fig. 3.2 Kim's argument

---

[10] Apart from the cases loved by metaphysicians, such as when two bullets hit a person at exactly the same moment and each one would be sufficient to kill them. Such cases do not concern us here.

3.5   Six Premises

Let's now unpack the argument of the previous section. It can be restated as a list of six premises, a maximum of five of which can hold without contradiction.[11] We'll see in subsequent sections that the choice of rejected premise corresponds to a view of the world. The six premises are dependence, reality, efficacy, distinctness, physical causal closure and single cause. Here are brief definitions.

Dependence: High level features depend on the features of their microphysical base. The simplest way to think about this is that whenever the microphysical features appear, so too do the high level features.[12]

Reality: Higher level features are real.

Efficacy: Higher level features have causal powers.

Distinctness: The features of higher levels are distinct from those of their microphysical base.

Physical Causal Closure: All physical effects are fully determined by prior physical occurrences.[13] To explain things at the lowest level, physics is all you need.

Single Cause: no event can have more than one sufficient cause occurring at any given time.[14]

It's worth dwelling a bit longer on the last of these. First, note that it presumes we don't have to worry about genuine cases of simultaneous causation such as that described in footnote 10. Secondly, it seems to clash with commonsense usage which regularly describes things having multiple causes, for example "the roof fell down because it was old and there was a storm". But the definition talks about a single cause "at any given time". Common usage is usually about temporal ordering in a causal chain, "being old" comes before "a storm". If it was the other way round, the roof would not have fallen

---

[11] This approach is borrowed from Wilson (2021).

[12] Noordhof (2019) counts no less than eleven distinct meanings of "dependence". The notions of dependence is also closely related to that of supervenience, see McLaughlin and Bennett (2021).

[13] This comes from Papineau (2001). Gibb (2015) cites eight other formulations.

[14] This formulation is from Kim (2007), p. 42.

down. When the roof falls down, being old is a state, the only cause occurring at that time is the storm.

We can now restate the argument of the previous section in terms of these premises. Again let's take higher level feature H which, by *Dependence*, has some microphysical base P. Given *Efficacy*, H can cause another higher level feature H* which, by *Dependence*, has some microphysical base P*. By *Physical Causal Closure* P* is caused by P and by *Dependence* P* causes H*. This is shown in Fig. 3.3.

Then using *Reality* and *Distinctness* (show on the diagram by the different colour for P and H), it is easy to see that H* is caused by both P* and H, so violating *Single Cause*. In the case where instead of H causing H* directly it does so by first causing P*, *Single Cause* is violated for P*.

Kim's argument is that if the first five of the premises are true, the sixth cannot be. More generally, no more than five out of the six premises can be true. Which premise you choose to reject corresponds to a distinct metaphysical position as shown in Table 3.1. Let's investigate them in turn.

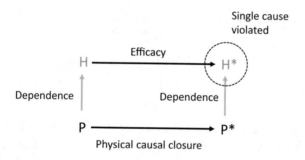

Fig. 3.3 Overdetermination

Table 3.1 Six metaphysical positions

Premise rejected	Position
Dependence	Dualism
Reality	Eliminativism
Efficacy	Epiphenomenalism
Distinctness	Reductionism
Physical causal closure	Strong emergence
Single cause	Weak emergence

3.6   It's Magic: Dualism

If we reject *Dependence*, then higher level features are independent of their microphysical bases. This means there is no causal chain going from P to H* and *Single Cause* holds. The higher level floats free of the physical: this is dualism.

Figure 3.4 is an illustration. Causation happens at the higher level. If you want the higher level to have physical effects, *Dependence* must be replaced with *downward causation*.[15] The arrows go from H to P*, from the higher level to the lower level. The microphysical base somehow acquiesces to whatever the higher level demands. This violates physical causal closure, shown on the diagram by the broken arrow of causation going from P to P*. Where the spirit world is concerned, the rules of physics are suspended. This is the first time we've seen downward causation, where the higher level causes features of the microphysical base. It will not be the last.

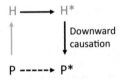

Fig. 3.4   Dualism

To a plain vanilla dualist, instead of being layered, the world is bifurcated. On the one hand, the spirit domain obeys its own rules. On the other, the microphysical base just does what it is told. A different form of dualism is panpsychism, which says spirit, mind or whatever is a fundamental property of matter.

Some form of dualism seems to be wired deep into our intuitions about how the world works. Today, dualist arguments are mostly confined to discussions of consciousness, but it's not all that long since dualist explanations were ubiquitous, just think of weather gods, water spirits or demonic possession. More on this in the next chapter.

---

[15] The term downward causation was first used in 1974 by Donald Campbell in a discussion of emergence in biological evolution, Campbell (1974).

Box 3.1 Reality and illusion

To be an illusion is to be unreal. Hence to understand what is illusion, we need to understand what is real. There are many philosophical definitions of reality. One, known as the Eleatic Doctrine, defines reality as having causal power. Here is the original passage:

> …a thing genuinely is if it has some capacity, of whatever sort, either to act on another thing, of whatever nature, or to be acted on, even to the slightest degree and by the most trivial of things, and even if it is just the once. That is, what marks off the things that are as being.[16]

If all causation happens at the lowest level, this then implies that everything else is unreal. I will argue that in this sense, emergence is an illusion.

Another definition of reality is mind-independence. Chapter 8 shows that the levels and special sciences I described in the first section of this chapter are not properties of the world but features of our cognitive limitations. In this sense too, emergent phenomena are illusions.

The Enlightenment philosopher John Locke defines secondary qualities as representing the causal power of things in the world to produce effects in us. We'll see in Chap. 13 that this is the most one can hope for in an eliminativist account. Things, physical systems, interact with the human brain, another physical system, and in doing so change it. Every detail of the interaction is described by fundamental physics.

3.7 It's an Illusion: Eliminativism

Rejecting *Reality* means that higher level features are illusions. Since illusions cannot be causes, *Single Cause* holds. Such a position is called eliminativism. We used to explain natural phenomena in dualist terms, invoking spirits, demons or gods. Science has eliminated these concepts, showing them to be illusions, no more that projections of our intuitions onto the world. Eliminativism says that a similar process will apply to all higher level phenomena. The only level that has reality is the lowest, physics. Just as we no longer explain things in terms demons, so, once we know the physics (and have enough computing power) we will stop explaining things in the terms of chemistry, biology or psychology. Instead, all explanations will be at the lowest level.

---

[16] Plato and Rowe (2015), p. 145.

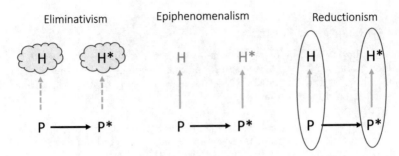

Fig. 3.5 It's all physics

This is shown in the left panel of Fig. 3.5. I've shown the higher level features in clouds to represent their existing only in minds. Eliminativism means the world we experience is radically different from the world as it really is. Part III of the book argues for an eliminative stance towards emergent (and indeed all) phenomena.

3.8   It's Almost an Illusion: Epiphenomenalism

Rejecting *Efficacy* means that higher level features exist but have no causal powers. Then *Single Cause* can hold. In philosophical terms, higher level features are epiphenomenal. Philosophers use the concept to construct arguments about the nature of *Dependence*. Some talk of *epiphenomenal ectoplasm*,[17] others of *idlers*, fundamental properties of the world which play no active role in it.[18]

This is shown in the middle panel of Fig. 3.5. There is now no arrow going from H to H*. While they exist, higher level features are like the ghosts of ancient Greece, wandering the world without being able to change anything. It's hard to find examples and I suspect this is implicit in the definition. It's even harder to find an example that distinguishes between epiphenomenalism and eliminativism. This should not be surprising as *Reality* is sometimes defined as having causal powers (see Box 3.1).

It seems to me that consciousness is perhaps the only phenomenon that may be epiphenomenal. One model of consciousness is that it is something like the display of a computer, just a useful representation of the internal

---

[17] Stoljar (2010), Sect. 7.5.
[18] Lewis (2008).

state of a system but without any causal powers in its own right. It's just a first-person phenomenon, with no objective existence, making no difference anywhere. Though it could be argued that if it were truly epiphenomenal, we could not know of its existence. Instead consciousness has at least enough causal power to make many people discuss it endlessly. From now on, I am going to treat epiphenomanlism as equivalent to eliminativism.

3.9   It's All Physics: Reductionism

Rejecting *Distinctness* implies that higher level features have no properties over and above those of their material base. Since H and P are the same, *Single Cause* holds. This is reductionism, higher level features are real, but can be reduced to physics. The right panel of Fig. 3.5 illustrates this. Note that H and P are shown in the same colour since there is no difference between them.

A long-lasting debate in the philosophy of science asks what, exactly, constitutes reduction. This passage, from the introduction to a review article, captures the way in which I use the term:

> ...all varieties [of reduction] share "nothing-but"-ism: a reduction shows that the reduced kind (whatever kind it might be) is thereby "nothing but" the reducing; no reduced content is left out or over. This is the feature that sets reduction in opposition to emergence.[19]

Reductionism upsets people and even causes outrage. In general, it has had a terrible press to the extent that it is often found prefixed by the word 'crass'. This seems to me undeserved. I will return to this in Chap. 13.

3.10   Strong Emergence

Rejecting *Physical Causal Closure* implies that higher level features have novel causal powers over and above those of their microphysical base. This is strong emergence, illustrated in Fig. 3.6. It can happen in two ways. In the left panel, H causes H* directly, and then H* causes the low level P*. In the right panel, H causes P* directly and then H* is automatically instantiated by *Dependence*.

---

[19] Bickle (2019).

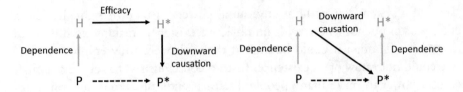

Fig. 3.6 Strong emergence

In either case, there is downward causation, the higher level has causal power distinct from the lower level. This implies that P is no longer a sufficient cause of P* and *Physical Causal Closure* is violated. This is indicated on the figure by the broken arrow. Lower levels constitute higher levels by *Dependence*. But higher levels also determine the behaviour of lower levels by some "mysterious brute determination".[20] This two-way causal relation is characteristic of strong emergence.

3.11 Weak Emergence

The final strategy is to accept the first five premises and reject *Single Cause*. Weak emergence argues that all causation happens at the lowest level but higher level features can have *fewer* causal powers than the lowest level. Since they have a different set of causal powers, higher level features are then metaphysically distinct from their microphysical base.

To explain this, let me adapt an example from Fodor.[21] Take a higher level object such as a coke machine. The special science of coke machines has three laws:

First Law: put the right coin in the machine, and you get a coke.

Second law: put the wrong coin in the machine, and it is returned.

Third law: when the machine is empty, all coins are returned.

These laws are valid irrespective of whether the coke machine is mechanical, electrical, electronic or, indeed, if it is controlled by a person sitting inside it. Everything that goes on inside the coke machine is a consequence of its microphysical base. But to use a coke machine, you just need to know

---

[20] Yates (2017).
[21] Fodor (1981).

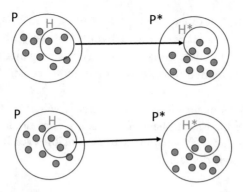

Fig. 3.7 Weak emergence

the three laws. Almost all of the causal powers of the machine's microphysical base are irrelevant. Only a subset of these powers captured by the three laws are needed to fully understand its operation.

This is the characteristic of weak emergence. Since they have a distinct set of causal powers, the higher level features of a coke machine are different from their microphysical base so cannot be reduced to it. This is illustrated in Fig. 3.7. The large circles correspond to the microphysical bases P and P*. Casual powers are represented by dots. High level features, with a subset of causal powers, are shown by the smaller circles.

In the top panel, H causes H* directly; in the bottom panel H causes P* which then by *Dependence* instantiates H*. In both cases, there is downward causation and H* has two causes so *Single Cause* is violated. But this is benign:, "…it is no more problematic than in cases where, e.g., both a plane and its wheels are causes of a runway's being touched".[22]

Strong emergence argues that higher level features have **more** causal powers than their microphysical base so are distinct from it. Weak emergence argues that higher level features have **fewer** causal powers than their base, so again must be distinct from it. Whereas strong emergence contradicts physics, weak emergence doesn't.

3.12 Epistemic Emergence

These are metaphysical definition of emergence in terms of causal powers. They imply that high level features are just as ontologically real as their microphysical base. Scientists try to explain features of the world. Generally, they

---

[22] Wilson (2021), p. 70.

don't worry too much about metaphysics, ontology or causation. Of the definitions listed in Appendix A.1, only a handful use any of these terms. The philosopher David Chalmers gives definitions of emergence which have more practical import.[23] His definition of strong emergence is:

> ...a high level phenomenon is strongly emergent with respect to a low level domain when the high level phenomenon arises from the low level domain, truths concerning that phenomenon are **not deducible even in principle** from truths in the low level domain.

and of weak emergence.

> ...when the high level phenomenon arises from the low level domain, but truths concerning that phenomenon are **unexpected** given the principles governing the low level domain.

These definitions are in terms of the state of our knowledge, whether of mathematics ("deducible even in principle") or the state of the world ("unexpected"). This is epistemic emergence. What is the relation with metaphysical emergence? Wilson argues that there is none:

> ... failures of predictability or derivability [do not] have any clear metaphysical consequences for whether there are distinct and distinctively efficacious higher level entities[24]

The issue of whether emergence is epistemic or ontological will be a recurring theme of Part II.

3.13 Physicalism

If you are a physicalist, you think everything is, ultimately, physics. There are broadly three kinds of physicalism:

Reductive physicalism: The lowest level of the hierarchy is physics. In its reductive variety, physicalism says that all causation happens at this level. The causal powers of the higher levels drain away to this lowest level. If,

---

[23] Chalmers (2006).
[24] Wilson (2021), p. 13.

as suggested in the previous section, there is no lowest level, then causation drains away into a "bottomless pit".[25] Higher level features exist but are identical to lower level features.

Eliminative physicalism: This takes reductive physicalism one step further by saying higher level features are illusions. Here is how Jaegwon Kim distinguishes the two:

> There is an honest difference between elimination and conservative reduction. Phlogiston was eliminated, not reduced; temperature and heat were reduced, not eliminated. Witches were eliminated, not reduced; the gene has been reduced, not eliminated.[26]

I do not agree with Kim and will argue in Chap. 13 that reductive physicalism necessarily collapses into eliminativism.

Non-reductive physicalism: this is another name for weak emergence. Higher levels have no causal powers distinct from the base, but since they only have a subset of these powers a meaningful notion of causality is kept at higher levels. Reduction is not possible. More is different, but in a way consistent with physics.

There is something odd about non-reductive physicalism. It tries to slip around the causal exclusion argument by asserting the things can have multiple causes but in an inoffensive way. Alex Rosenberg captures this awkwardness with an argument[27] which can be paraphrased as:

1. Non-reductive physicalists claim that facts about special sciences are physical but cannot be explained by physics.[28]
2. This means there are two types of fundamental things in the world, microphysics and special science facts.
3. If these facts are physical, then we are back to physicalism
4. If there are not physical, then we need to reject physicalism.

Without going any further, this seems to rule out non-reductive physicalism. If you are convinced by Rosenberg's argument, there is no need to read Chaps. 6–9 which discuss various forms of weak emergence.

---

[25] Schaffer (2003).

[26] Kim (2007), p. 160.

[27] Rosenberg (2006), p. 7.

[28] For a contrasting view, see Wilson (2010), Sect. 5.1.1.

An aside on terminology. Philosophers tend to use physicalism in preference to materialism. This is partly because the content of modern physics is both matter and fields. Partly because the term materialism comes with baggage, either from the Marxist tradition or the commonsense meaning of a way of life devoted to material accumulation. And partly because

> …materialism's modern descendants have—understandably—lost their metaphysical nerve. No longer trying to limit the matter of physics a priori, they now take a more subservient attitude: the empirical world, they claim, contains just what a true complete physical science would say it contains.[29]

I am going to follow convention and stick with physicalism despite my far preferring materialism on aesthetic grounds. I was delighted to find I shared this taste with Ludwig Wittgenstein, who in 1932 wrote

> It is not true that I have not dealt with the question of 'physicalism' (albeit not under this – dreadful – name) and with the same brevity with which the entire Tractatus is written[30]

This might all sound rather dull. But these are radical positions. On the one hand, non-reductive physicalism implies that while higher level features are nothing but physical features, they are not just physical features. This holds the promise of being able to be a physicalist while preserving deep intuitions about the reality of higher level features, steering a middle way between the violence of reductionism and the superstition of dualism. On the other, taking eliminative physicalism seriously, which is what Part III of this book does, means we need to treat everything we know apart from physics as suspect:

> As a result, even on an optimistic assessment, practically all causal explanations we de facto use will turn out to be false (since we give extremely few—if any—such explanations in terms of phenomena at a fundamental microphysical level, if there is any such level at all).[31]

You don't get much more radical than that.

---

[29] Crane and Mellor (1990).
[30] Quoted in Stern (2007).
[31] Mayr (2017).

3.14   The Limits of Physics 1

Physics, like science in general, is work in progress. On the face of it, this seems to pose a problem for physicalism. You may think that everything is physics, but what is physics? There are two possible answers. Either physics is understood to be current physics. Or it is taken to be some future Theory of Everything.

We can be pretty sure that current physics is incomplete and inaccurate. At it's heart are some profound and unresolved issues, among them the relation between quantum physics and general relativity at high energies and the nature of dark energy and matter. This means that on the one hand, if physicalism is based on current physics, it must be false. And on other, if it is based on some currently non-existent theory, we cannot know what it is. Imagine if the panpsychists are correct and a psychic realm exists. Imagine further that it proves amenable to scientific enquiry. Then we would need to extend physics to include psychophysical laws. There's no logical reason this could not be the case and it illustrates how unknowable is the form of future physics. This argument that physicalism is either false or unknowable has come to be known as Hempel's dilemma after the philosopher Carl Hempel.[32]

There have been various attempts to resolve the dilemma. The one I find most satisfactory sees physicalism as a dynamic process the meaning of which changes with scientific knowledge. Andrew Melnyk sums this up:

> …it is perfectly possible to endorse any scientific hypothesis, including therefore physicalism, and indeed to endorse it rationally, while acknowledging that it has only a very low probability of being true.[33]

Physicalism is just like any other scientific concept in that it is based on our current knowledge. That's the best we can ever do.

The domain of physics could also be limited by the existence of abstract entities. To the assertion that everything is physics, a philosopher might retort: what about numbers? What about the US Supreme Court? My view is that such things only exist if they are instantiated as a microphysical pattern in someone's brain (or in some other entity's cognitive apparatus). Do numbers exist in some Platonic world that pre-exists entities that can use them? I've no idea. Nor can I imagine that such a view has any testable consequences.

---

[32] Hempel (1969) first states the dilemma in terms of biology.
[33] Melnyk (2003), p. 14.

3.15 Definitions

This chapter might have given the impression that the definition of emergence is a straightforward affair. The 75 definitions in Appendix A.1 suggest this is not the case. To try to make some sense of these, I used various tricks, including the word cloud of Fig. 3.8. However nothing gave me any useful insights and I shall leave this interesting exercise to others.

In the meantime, there are two basic strategies one can take towards the mass of definitions. The first says we should be pluralistic and not worry too much. Here is Michael Silberstein

> Different cases require different conceptions of emergence. It is absurd for philosophers to try and argue that any particular conception of emergence is inherently superior across the board.[34]

The other is to conclude that if, after decades of study, a concept remains such a confused mess, it is probably best discarded. While this captures my position on emergence, Part II studies several definitions and investigates to what extent they are satisfactory. To do this, we first need to ask what makes a good definition.

Apart from some underlying features, such as relevance, clarity and consistency, the important property seems to me that of splitting up the world in

Fig. 3.8 Definitions

---

[34] Silberstein (2012).

a useful way. A definition of X should explain why some things are X and some things are not X. Such a definition can be fuzzy. Indeed, after Wittgenstein, it is hard to argue that things have precise boundaries. But to be useful, a definition still should imply that some things mostly resemble each other and some things mostly don't.

This is what I'd expect from a useful definition of emergence. I'd like to be able to go through the examples of the previous chapter and say: yes, that one is emergent because of these properties; that one isn't because of these properties; and this one, well, that's a bit of a grey area, these properties fit the definition while these ones don't.

Just to be clear, I am not concerned with casual uses of the term. Complex behaviour emerges from simple rules can mean no more than simple rules lead to or result in complex behaviour.

3.16 Discussion

A different way to look at the six positions presented is this: what do they imply about the knowledge we need to understand the world? What do you need to know to know everything that matters?

For reductionism, eliminativism and epiphenomenalism, the answer is straightforward. Microphysics is the only thing that has causal power, so once you know the microphysics, you know everything that matters. The higher level follows naturally. For the other three, physics is not enough. For dualism, you also need to know the rules of the spirit world. For strong emergence, you need to know the laws governing the higher level and a description of how physics is coerced by downward causation. For weak emergence, you need the special science to tell you what subset of physical features are relevant.

This neatly leads to the key motivation of this book. If one of the first three positions is correct, physics fixes all the facts. If there is dualism or either form of emergence, physics is not enough. That's why the question of the reality of emergence matters.

3.17 Further Reading

The main source this chapter is the clear and enlightening Wilson (2021). For a different approach to metaphysics, see Ladyman and Ross (2009).

A history of the concept of emergence is in Blitz (1988); Caston (1997) reviews its role in ancient Greek thought. For discussions of levels from a philosophical perspective, see Kim (2002) and Tahko (2023). An interesting comparison of the way the idea of levels is used by philosophers and physicists is in Rueger and McGivern (2010). For more on the idea of a fundamental level, see French (2022).

A clear presentation of dualism is in the Stanford Encyclopaedia of Philosophy, Robinson (2023), and Seager (2018) discusses panpsychism. Two useful books on physicalism are Stoljar (2010) and Melnyk (2003). For a discussion of physical causal closure, see Gibb (2015). For an overview of reductionism, see Bickle (2019) and for a history of reductionism by a philosopher of science, Tahko (2021).

A good place to start on eliminativism is the classic Churchland (1981); the Stanford Encyclopaedia of Philosophy, Ramsey (2022), gives a thorough overview. Discussion of Hempel's dilemma and possible resolutions can be found in Stoljar (2010), Chap. 5 and Stoljar (2023), Sect. 4.

More suggestions for reading can be found at www.TheMaterialWorld.net.

References

Aristotle, Reeve CDC (2016) Metaphysics. Hackett Publishing Company, Indianapolis ; Cambridge

Bickle J (2019) Reduction. In: Gibb SC (ed) The Routledge handbook of emergence. Routledge, New York

Blitz D (1988) Emergent evolution. McGill University, Montreal

Campbell DT (1974) 'Downward Causation' in Hierarchically Organised Biological Systems. In: Ayala FJ, Dobzhansky T (eds) Studies in the Philosophy of Biology. Macmillan Education UK, London, pp 179–186

Caston V (1997) Epiphenomenalisms, Ancient and Modern. The Philosophical Review 106:309. https://doi.org/10.2307/2998397

Chalmers DJ (2006) Strong and Weak Emergence. In: Davies P, Clayton P (eds) The Re-Emergence of Emergence: The Emergentist Hypothesis From Science to Religion. Oxford University Press

Churchland PM (1981) Eliminative Materialism and the Propositional Attitudes. The Journal of Philosophy 78:67. https://doi.org/10.2307/2025900

Crane T, Mellor DH (1990) There is No Question of Physicalism. Mind XCIX:185–206. https://doi.org/10.1093/mind/XCIX.394.185

Fodor JA (1981) The Mind-Body Problem. Scientific American 244:114–123. https://doi.org/10.1038/scientificamerican0181-114

Fraser D (2022) Particles in quantum field theory. In: The Routledge companion to philosophy of physics. Routledge, New York

French S (2022) Fundamentality. In: The Routledge companion to philosophy of physics. Routledge, New York

Gibb S (2015) The Causal Closure Principle. Phil Quart 65:626–647. https://doi.org/10.1093/pq/pqv030

Hempel C (1969) Reduction: Ontological and Linguistic Facets. In: Morgenbesser MPSSW (ed) Philosophy, Science, and Method: Essays in Honor of Ernest Nagel. St Martin's Press

Kim J (2002) The Layered Model: Metaphysical Considerations. Philosophical Explorations 5:2–20. https://doi.org/10.1080/10002002018538719

Kim J (2007) Physicalism, or Something Near Enough, Princeton University Press, Princeton (N.J.)

Ladyman J, Ross D (2009) Every thing must go: metaphysics naturalized. Oxford University Press, Oxford

Lewis D (2008) Ramseyan Humility. In: Braddon-Mitchell, David, Nola, Robert (eds) Conceptual Analysis and Philosophical Naturalism. The MIT Press

Mayr E (2017) Powers and Downward Causation. In: Paolini Paoletti M (ed) Philosophical and scientific perspectives on downward causation. Routledge, Taylor & Francis Group, New York

McLaughlin B, Bennett K (2021) Supervenience. The Stanford Encyclopedia of Philosophy. https://plato.stanford.edu/entries/supervenience/

Melnyk A (2003) A physicalist manifesto: thoroughly modern materialism. Cambridge University Press, Cambridge

Newton I (2012) Opticks: or, a treatise of the reflections, refractions, inflections, & colours of light. Dover Publications, Inc, Mineola, New York

Noordhof P (2019) Dependence. In: Gibb SC (ed) The Routledge handbook of emergence. Routledge, New York

Papineau D (2001) The Rise of Physicalism. In: Gillett C, Loewer B (eds) Physicalism and its Discontents. Cambridge University Press, pp 3–36

Pennac D (1995) Comme un roman. Gallimard, Paris

Plato, Rowe C (2015) Theaetetus and Sophist. Cambridge University Press, Cambridge

Popper KR (1995) Objective knowledge: an evolutionary approach. Clarendon Press, Oxford

Ramsey W (2022) Eliminative Materialism. The Stanford Encyclopedia of Philosophy. https://plato.stanford.edu/entries/materialism-eliminative/

Robinson H (2023) Dualism. The Stanford Encyclopedia of Philosophy. https://plato.stanford.edu/entries/dualism/

Rosenberg A (2006) Darwinian reductionism, or, how to stop worrying and love molecular biology. University of Chicago Press, Chicago

Rueger A, McGivern P (2010) Hierarchies and levels of reality. Synthese 176:379–397. https://doi.org/10.1007/s11229-009-9572-2

Schaffer J (2003) Is There a Fundamental Level? Noûs 37:498–517. https://doi.org/10.1111/1468-0068.00448

Seager W (2018) Idealism, Panpsychism, and Emergentism. In: Gennaro RJ (ed) The Routledge handbook of consciousness, Routledge, New York

Silberstein M (2012) Emergence and reduction in context: Philosophy of science and/or analytic metaphysics. Metascience 21:627–642. https://doi.org/10.1007/s11016-012-9671-4

Stern D (2007) Wittgenstein, the Vienna Circle, and Physicalism: A Reassessment. In: Richardson A, Uebel T (eds) The Cambridge Companion to Logical Empiricism. Cambridge University Press, pp 305–331

Stoljar D (2010) Physicalism. Routledge, London ; New York

Stoljar D (2023) Physicalism. The Stanford Encyclopedia of Philosophy. https://plato.stanford.edu/entries/physicalism/

Tahko TE (2021) Unity of science. Cambridge University Press, Cambridge

Tahko TE (2023) Fundamentality. The Stanford Encyclopedia of Philosophy. https://plato.stanford.edu/entries/fundamentality/

Tegmark M (2008) The Mathematical Universe. Found Phys 38:101–150. https://doi.org/10.1007/s10701-007-9186-9

Wilson JM (2010) Non-reductive Physicalism and Degrees of Freedom. The British Journal for the Philosophy of Science 61:279–311. https://doi.org/10.1093/bjps/axp040

Wilson JM (2021) Metaphysical Emergence. Oxford University Press, Oxford

Yates D (2017) Demystifying Emergence. Ergo, an Open Access Journal of Philosophy 3:. https://doi.org/10.3998/ergo.12405314.0003.031

4

The View from the Cave

Summary To understand why we interpret phenomena as emergent, we need to know something about the interpreting apparatus, the human cognitive system. The constraints faced by cognitive evolution lead to the development of fast and frugal heuristics which are about enhancing evolutionary fitness not discovering the nature of the world. Perhaps the central aspect of humanity's recent cognitive evolution is social learning. This, along with cognitive dispositions we share with our ancestors, forms the base for the system of distributed cognition which is the scientific project. Are there limits to this project? The question leads to the distinction between imaginative and representational understanding which will be central to the discussion of emergence.

Are emergent properties features of the world? Or are they features of what we know about the world? The previous chapter introduced the distinction between ontological and epistemic emergence. I think this is a false distinction. It is an illusion to imagine that our thoughts give us direct access to the nature of reality. Ontology is always epistemic. Metaphysical certainty only lasts until we hear a better argument and think "Well, of course".

Debates over the this are a constant theme in Western philosophy. Plato's metaphor of the cave highlights the difference.[1] Ordinary people only see

---

[1] Plato et al. (2007), bk. VII.

© The Author(s), under exclusive license to Springer Nature
Switzerland AG 2025
L. Graham, *Physics Fixes All the Facts*, The Frontiers Collection,
https://doi.org/10.1007/978-3-031-69288-8_4

shadows on a wall without being able to see what created them. Philosophers can step into the light and see the nature of reality. It's not hard to find traces of this condescending attitude throughout philosophy, mere epistemic emergence versus the proper ontological version. A more recent version of the distinction is found in Immanuel Kant's assertion that we can only experience the world through categories that are properties of our minds and not of the world. All we ever have is the phenomena. Unlike the cave-dwellers who can at least study philosophy, Kant argues that we have no way of accessing the intrinsic properties of objects that are behind the phenomena.

None of this has ever been much of an issue for science. Why care about intrinsic properties since, by definition, they can never affect us? Scientific explanations are only descriptions of the world. If you take a child's approach of responding "Why?" to every answer, at some point a scientist will have to say, "That's just the way it is" or "If it wasn't this way, there'd be no possibility of creatures with the capacity to ask why".

Despite this, it is science, not philosophy, which allows us to investigate the world outside the cave. Here is neuroscientist Thomas Metzinger describing the neurophenomenological cave that he calls the ego tunnel:

> What we see and hear, or what we feel and smell and taste, is only a small fraction of what actually exists out there. Our conscious model of reality is a low-dimensional projection of the inconceivably richer physical reality surrounding and sustaining us. Our sensory organs are limited: They evolved for reasons of survival, not for depicting the enormous wealth and richness of reality in all its unfathomable depth.[2]

Compare our ability to discriminate between colours with the richness of the spectrum of visible light or with the full spectrum of electromagnetic radiation from long-wave radio to gamma rays. Repeat the exercise for the other senses. Then for senses that we don't have but other animals do: echolocation, sensitivity to electric fields or the polarization of light. Then take a trip into science fiction and imagine the senses that are possible. This is what I take Metzinger to mean when he describes the world we experience as a low dimensional projection.

But perception is just one part of the human cognitive apparatus. To investigate the others, we need to start by studying the capacities of our brains, then turn to what often seems the biggest miracle of all: how a cognitive apparatus that evolved to enhance our survival on the African savannah allows us to do science.

---

[2] Metzinger (2009).

4.1 Constraints

Cognitive abilities are one source of selective advantage. The human cognitive system is the result of a long evolutionary process, stretching back to the earliest bacteria that were capable of responding to their environment. To understand its nature, we need to understand the constraints faced by this evolutionary process. In a paper from 2020, computational cognitive scientist Tom Griffiths argues that these constraints can be divided into three categories.[3]

The first type of constraint is about time. The most basic limit is imposed by the need to survive. If humans needed thousands of examples to reliably identify a predator, they wouldn't live long. Griffiths writes:

> ...the capacity to learn from limited data is comparable with the ability of baby gazelles to run shortly after birth, a consequence of the limited time available to build up the requisite skills for survival.[4]

Another limit comes from the trade-off between time spent learning and time spent using what is learnt. Then there is the limit of human lifetimes. Our innate models of the world consist of genetically encoded rules which are then tuned by experience. The tighter these time constraints, the more the balance is shifted towards hardwiring and away from tuning.

The second type of constraint is about computational resources. The brain has limited processing power and memory capacity. Much of this is devoted to background activities such as the perceptual and motor systems so is not available for learning. Griffiths argues that these limits mean humans have become good at splitting problems into parts. This allows them to be solved in series. Along the way, previously learnt solutions can be applied to the subproblems, further reducing the resources needed.

The third type of constraint is due to limited communication. The other constraints would be looser if problems could be effortlessly shared among individuals. Being eaten by a predator would not matter (at least in learning terms) if the information so acquired could be transferred to the rest of the group. Similarly, the computational resources constraint would matter less if problems could be split up into smaller chunks, spread across individuals to be solved then aggregated. But both of these would require an efficient way of transferring information between individuals and evolution has not provided us with brain-to-brain interfaces.

---

[3] Griffiths (2020).
[4] Griffiths (2020).

What are the implications of these constraints? Those of time and computational resources lead to our innate model of the world being made up of *fast and frugal heuristics*. The absence of direct communication results in language and cultural evolution. I will discuss these in the next sections.

4.2   Fast and Frugal

Imagine you want to design a robot to catch a ball. Putting the problems of measurement and movement to one side, you need to write some code to predict the ball's trajectory. This is not an easy exercise, requiring an awful lot of fluid dynamics and some serious computing power. When you've done that, you can predict where the ball is going to be at any point in time and, if you've done the calculation quickly enough, you can send the robot in a straight line at a speed that will place it under the ball. Putting all this together is a formidable technical challenge and the first robot catcher was demonstrated in 2011, working in laboratory conditions.[5]

But many humans can do it pretty well, with a bit of practice, and some dogs do it even better. This doesn't involve calculating the trajectory of a ball or predicting where it will fall. Instead, people use a simple rule: keep your eye on the ball and run in a direction such that the angle of the ball is constant.[6] Following the rule will take you along a curved path to where the ball lands. This is illustrated in Fig. 4.1. The ball follows a parabola. The fielder's position is shown by the baseball cap and their gaze by the arrows. The result is that the fielder runs on a curved path at varying speeds. If you turn this into a quantitative model, its predictions match actual behaviour well.

No knowledge of physics, complex measurements or calculations are involved. All that is needed is to keep your eye on the ball and some motor coordination. This is often used as the canonical example of a fast and frugal heuristic. Fast in the sense that it can solve the problem quickly, just fix your gaze on the ball start to run. Frugal in the sense that it requires little information, just the angle of the gaze, and exploits the existing capacities of the perceptual and motor systems.

Such heuristics are context-specific. We should expect them to work pretty well in the environment they developed in. But if we change the environment they will work less well, or not at all. Even if you are proficient at ball catching, you need to relearn how to catch a frisbee or a shuttlecock. These

---

[5] Bauml et al. (2011). A video of the robot in action can be found at www.TheMaterialWorld.net.
[6] It's a bit more complex than that, see McLeod et al. (2003).

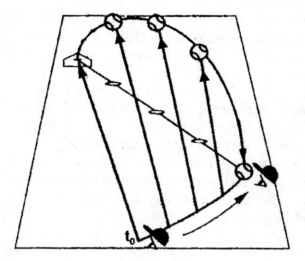

Fig. 4.1 Catch![7]

limits can be seen in other creatures. Birds fly into windows; they evolved in environments without vertical reflective surfaces. Frogs allow themselves (supposedly) to be slowly boiled to death; they evolved in environments without malicious biologists.

We are equipped with a rich set of heuristics, or intuitions, about the world and the entities in it. To describe them, philosophers prefix the word *folk* to their domain of application giving folk psychology, folk biology or folk physics. I prefer using *commonsense* as a prefix. In evolutionary terms, the point of such intuitions is the same as that of everything else: to help us survive and reproduce in our ancestral environment.

Let me give some examples, starting with commonsense physics. Imagine you point a high-powered rifle parallel to the ground. Then at the same instant you pull the trigger, you drop a bullet from the same height as the barrel. Which bullet will hit the ground first? Many people will answer that the dropped bullet will, because the forward-moving impetus of the other will keep it in the air longer.[8]

This idea of impetus is part of commonsense physics and was a central component of prescientific theories of motion. One of the key moments in the transition to modern science was the demonstration that it is incorrect. Drop a cannon ball from the mast of a ship at rest, and it will fall at the base of the mast. However, if the ship is sailing at speed it seems obvious, and

---

[7] Copyright © 2002, American Psychological Association. Reproduced with permission. Shaffer and McBeath (2002).

[8] In fact, the dropped bullet will indeed hit the ground first, but due to the curvature of the earth.

the theory of impetus implies, that the canon ball will land further towards the back of the boat. In 1632, Galileo argued this was wrong and that the canon ball lands in the same position irrespective of the speed of the ship. Experiments soon confirmed this.[9]

Three and a half centuries later, a group of psychologists presented undergraduates with a version of Fig. 4.2 and asked them to pick which trajectory the ball would take. 62% of all subjects, and 40% of those having studied some physics, said the ball would fall straight down.[10]

Let's now turn to commonsense statistics. The Gambler's Fallacy is the belief that, in a game of chance, a number which has not turned up recently is more likely to be drawn in the future. If we repeatedly toss a coin and get a string of heads, we expect that the next toss is more likely to be tails. When we study statistics we learn that, as long as the game is fair, draws are independent of one another. Another example is the widespread perception that the digits of π are random. We shall see in the next chapter that they are anything but random and can be generated by an algorithm shorter than this sentence.

Then commonsense psychology. This involves treating entities as agents free to pursue their own desires and beliefs, interpreting their behaviour in terms of mental states such as hunger, anger or fear. You might respond that this is a perfectly reasonable way of seeing the world; I'll return to this question in Chap. 13. But throughout history people have not only applied it to

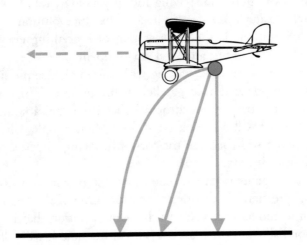

Fig. 4.2 A falling ball

---

[9] A video of one such experiment can be found at www.TheMaterialWorld.net.
[10] McCloskey et al. (1983).

other humans but to a whole range of natural objects and phenomena. We still curse tools for not cooperating with us. We talk to our domestic appliances. Anyone who has ever watched a robot lawnmower at work will know how easy it is to attribute mental states to it.

Box 4.1. Fast and slow?

Human cognition is about more than commonsense intuition and its fast and frugal heuristics. We are proud of our ability to reason and to consciously deliberate. A school of thought in psychology called *dual process theory* argues that there are two distinct systems. The first fast, largely automatic and extensively shared with other animals. The second slow, under conscious control (whatever that may mean) and unique to humans. Cognitive scientist Danny Kahneman (who shared the Nobel Memorial Prize in Economic Sciences, 2002) brought this distinction to popular notice with his 2012 book "Thinking fast and slow".[11] There is an interesting debate over whether this distinction is valid, but let's put this aside for the moment and focus on the constraints facing the slow process. In the absence of a good neurological understanding of what conscious thought is, it is difficult to understand the cognitive resources it requires. An exception is the concept of working memory. The general consensus seems to be that we can attend to no more than 3 or 4 things at any one time. If this is the case, and the area is littered with open questions, that is a tight constraint indeed.

A whole academic industry studies these heuristics and calls them cognitive biases. It has been referred to as the "people are stupid" school of psychology.[12] Since the mid-1990s, Gerd Gigerenzer has argued that instead such heuristics are exquisitely well adapted to our ancestral environment. Psychologists, trying to be good scientists, go to great lengths to eliminate context from their experiments. This creates artificial environments in which commonsense rules are bound to fail.

Take the Gambler's Fallacy. Natural events are not independent or random, so that observations of the past do give us useful information about the future. If it has rained for the past week it is more likely to be sunny tomorrow. If you've seen your prey taking the same route for a number of days, it is reasonable to assume they will take that route tomorrow. Here's Steven Pinker:

An astute observer should commit the gambler's fallacy and try to predict the next occurrence of an event from its history so far, a kind of statistics called

---

[11] Kahneman (2012).
[12] Kihlstrom (2004).

time-series analysis. There is one exception: devices that are designed to deliver events independently of their history. What kind of device would do that? We call them gambling machines. Their reason for being is to foil an observer who likes to turn patterns into predictions.[13]

But however much statistics or science we learn, commonsense intuitions are persistent. If you bet on the lottery, would you choose the numbers from one to six? If you hold a cup of coffee out in front of you, can you accept that the only force involved is exerted by your arm, preventing the cup and its contents from following their energy-minimising path in curved spacetime? A body of evidence suggests that education doesn't replace our common-sense models but that the two coexist. Empirical work suggests that scientists actively inhibit their commonsense intuitions when solving problems.[14]

4.3   How is Science Possible?

Fast and frugal heuristics may be rational in a given context, but they are still a long way from scientific concepts. This should be no surprise. They capture what was salient to us in the specific context of our ancestral environment. Why should they give us access to the nature of the world? What's worse, they tend to get in the way of learning science; every scientific discovery from Galileo onward contradicts some commonsense worldview. So how on earth do we manage to do science?

We can think of this in terms of the constraints in Sect. 4.1. Scientific problems require more time and computational resources than any individual has available. Lack of communication between brains restricts the possibility to spread the problems across people or time. The answer of how we get round these constraints is complex, but these components seem essential:

1. Some general cognitive dispositions which are useful both in our ancestral environment and for science.
2. Social learning leading to cultural evolution.
3. The use of objects to extend our abilities.

Let's look at these in turn. Steven Mithen traces the cognitive develop-ment of humanity and makes a convincing case that the cognitive dispositions

---

[13] Pinker (1999), p. 346.
[14] Brault Foisy et al. (2015).

necessary for science arose independently across the course of human evolution.[15] These include detailed and extensive observation of the natural world; creativity in the sense of the generation of hypotheses which have the potential for falsification; a concern with causation; tool use and the accumulation of knowledge through time.

I would add to this list the centrality of cooperation in early societies. This allows a physical division of labour and, later on, the intellectual division of labour that underlies science. The final item in the list points to the possibility of social learning and cultural evolution. One implies the other. If a creature can learn from its fellows, it can also learn from previous generations allowing knowledge to be transmitted across time and become culture.

Cultural evolution is absolutely central to the human story. The title of Joseph Henrich's 2015 book[16] calls it "The Secret of our Success". Successful social learning leads to selection for social learning skills and the cognitive abilities that these require leading to what Henrich describes as a "culture—gene evolutionary ratchet". Interestingly, both he and Mithen downplay the role of language in this process arguing that it is more a product of cultural evolution than a cause and that much social learning takes place without it.

At some point in this process, humans begin to use objects to extend their cognitive skills. This could be as simple as using an arrangement of sticks to provide external memory. Given how tight constraints on working memory seem to be (see Box 4.1), even the simplest external recording system could be expected to vastly increase the power of reasoning. In general, physical objects extend the scope of human cognition by providing abilities to record, store, transmit and process information and extend perception. This is often known as *cognitive scaffolding* and includes the whole range of material aids to cognition and also abstract tools such as diagrams or mathematics.

Cultural learning and cognitive scaffolding vastly amplify the basic cognitive capacities we share with our ancestors. Daniel Dennett writes:

> …human brains have become equipped with addons, thinking tools by the thousands, that multiply our brains' cognitive powers by many orders of magnitude. Language, as we have seen, is the key invention, and it expands our individual cognitive powers by providing a medium for uniting them with all the cognitive powers of every clever human being who has ever thought. The smartest chimpanzee never gets to compare notes with other chimpanzees in her group, let alone the millions of chimpanzees who have gone before.[17]

---

[15] Mithen (2002).
[16] Henrich (2016).
[17] Dennett (2017), Chap. 5.

Genetically, I am mostly the same as one of my hunter-gatherer ancestors from 10,000 years ago. Setting aside the differences in our upbringing (care, nutrition etc.), I have broadly the same general cognitive abilities. Yet as I sit here, I have a laptop in front of me. Stored on it are books and articles which represent the knowledge of hundreds of thousands of people accumulated over thousands of years. If I can't find what I need on my hard drive, it will be somewhere on the internet. As well as knowledge, the laptop provides tools to organise it and extend my memory. Pencil and paper are replaced by a word processor, an app for planning tasks and managing ideas and another to categorise and sort references. Also on the laptop is a simulation package which allows me to construct models of physical systems. If I forget something, whether the precise source of an idea, the syntax of a programming command or a mathematical derivation, I can look online. If I get stuck, I can send an email to a friend or contact a specialist. It's breathtaking and makes me feel a tiny part of an immense web stretching across space and time.

You can think of the result as a form of distributed cognition. It allows the process of science to be spread across time, individuals and physical system. This eases the constraints I described in Sect. 4.1. This division of labour also allows specialisation and results in what Andy Clark and David Chalmers refer to as the *extended mind*.[18]

Such distributed cognition is found throughout nature. Groups of entities can solve problems far beyond the capacities of their individuals. Slime moulds, consisting of an aggregation of single-celled organisms, can solve the travelling salesperson (of which more in Box 5.2) and other network optimisation problems.[19] The group behaviour of insects has inspired a computational approach to optimisation.[20] The human scientific project involves more capable individuals, but it is hard to see a qualitative difference.

The distributed aspect of the scientific project is also central from a philosophy of science perspective. Karl Popper wrote that the objectivity of science rests on the fact that hypotheses can be intersubjectively tested. Helen Longino, in what in my opinion is the best book[21] on methodology since Paul Feyerabend, extends Popper's argument. Intersubjectivity is not much

---

[18] Clark and Chalmers (1998). There is an interesting link to Friedrich Hayek's concept of capitalism as an extended order of human cooperation, Hayek and Bartley (2000).

[19] Sun (2019).

[20] Falcón-Cardona et al. (2022).

[21] Longino (1990).

use if it is just that of affluent white men. Longino argues that what guarantees the objectivity of science is the diversity of the criticisms that hypotheses face and hence the diversity of scientists.

4.4   The Limits of Physics 2

James Joyce wrote that rats will never understand trigonometry.[22] How arrogant it seems to assume that our brains are not similarly closed to some things. The idea that there is some limit to what we can understand is known as *cognitive closure*. Since the mid-1990s, Colin McGinn has argued that we are cognitively closed to some things including, unsurprisingly, the nature of consciousness. This position has come to be known as the New Mysterianism.

To think about this, I'll adapt the approach taken by a 2020 article.[23] Cognitive closure is too broad a term. There are at least three senses in which we can understand something. The first is *experiential understanding*, the ability to subjectively experience something. In the passage I quoted earlier in this chapter, Metzinger argues that we are experientially closed to many things due to the makeup of our perceptual system.

The second is *imaginative understanding*, the capacity to make sense of something in terms of our commonsense notions. Much of science teaching is about finding analogies, images or stories which make concepts accessible. Examples abound, from comparing the quantum double-slit experiment to the behaviour of waves in water to explaining general relativity by balls rolling on rubber sheets. In the case of imaginative closure:

> …it is impossible for us to comprehend the relevant scientific theory describing that part of reality. No matter how hard we try, we just can't wrap our minds around it. Because of some species-specific limitation to our imagination, this part of reality will forever bewilder and baffle us.[24]

The third is *representational understanding* in terms of scientific concepts, maths, models or whatever. If our cognitive structure means we are representationally closed to some things, they will forever remain mysteries to us.

To see the distinction between the imaginative and representational understanding, let's take a couple of examples. The first is a 4-dimensional cube or

---

[22] Joyce (2001).
[23] Boudry et al. (2020).
[24] Boudry et al. (2020).

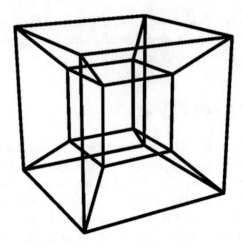

Fig. 4.3 A tesseract

tesseract. The mathematical properties of it are as well-understood as that of a 3D cube. Figure 4.3 shows a 2D projection of a tesseract. Online you can find videos showing how it can be unfolded into six 3D cubes in the same way a 3D cube can be unfolded into 6 2D squares.[25] We can represent it, but we can't imagine it.

The second example is quantum physics. The representational aspect, the mathematical formalism, is as well understood as anything and forms the basis of much of modern technology. Yet, famously, no one understands it. And no one says it better than Richard Feynman in the introduction to his popular book on Quantum Electrodynamics:

> What I am going to tell you about is what we teach our physics students in the third or fourth year of graduate school—and you think I'm going to explain it to you so you can understand it? No, you're not going to be able to understand it. Why, then, am I going to bother you with all this? Why are you going to sit here all this time, when you won't be able to understand what I am going to say? It is my task to convince you not to turn away because you don't understand it. You see, my physics students don't understand it either. That is because I don't understand it. Nobody does…. while I am describing to you how Nature works, you won't understand why Nature works that way. But you see, nobody understands that. I can't explain why Nature behaves in this peculiar way.[26]

[25] A video of an unfolding tesseract can be found at www.TheMaterialWorld.net.
[26] Feynman (2006), p. 24.

I take Feynman to be referring to what I called imaginative understanding. Given the limited scope of commonsense ideas, it is unsurprising that we fail to imaginatively understand quantum physics. The struggle to achieve imaginative understanding is a constant feature of scientific progress. Think of the initial reactions to the hypotheses that the earth moves, that heat is not a fluid and that we share a common ancestor with the dinosaurs.

This section started with the question of cognitive closure. We can reframe this as asking whether there are limits to representational understanding. And this makes it a completely different question, since it is no longer about the capacities of an individual brain but about those of the entire system of distributed cognition that is modern science. No one has direct representational access to more than a tiny corner of this system, but such access is always there if you need it.

Questions about cognitive closure then turn into a question about the limits of the scientific project. To argue that we have reached a limit, you would need a long period without scientific progress. On the contrary, it seems easy to make the case that there is unbounded scope for new tools, notably increasingly capable computers, to make new forms of cognitive scaffolding to represent the world in ways unimaginable to their human creators.

4.5   The Mind Projection Fallacy

The persistence of our commonsense intuitions and our inability to imagine scientific concepts have an important consequence. E.T. Jaynes called it the mind projection fallacy:

> ...we are all under an ego-driven temptation to project our private thoughts out onto the real world, by supposing that the creations of one's own imagination are real properties of Nature, or that one's own ignorance signifies some kind of indecision on the part of Nature.[27]

We fail to recognise our cognitive limitations and instead project them on to the world. This often involves mistaking our subjective experiences or judgements for objective properties. For example, observing the flatness of the world around you and assuming the earth is flat. Or treating the sweetness of sugar as a property of a sugar molecule rather than a complex interaction between the molecule and the human perceptual system.

---

[27] Jaynes (2010).

Philosophers seem particularly susceptible to this fallacy, with an unfortunate habit of treating their thought processes as objective data. This is implicit in the analogy of the cave: all it takes to find our way out into the sunlight is a look inside our minds. Coming up with other examples from the history of philosophy is an unfair sport. Aristotle maintained that spheres of different masses fall at different speeds. Kant claimed that Euclidean geometry and Newtonian physics were deducible a priori.

More recent examples abound. Here's one from an article on quantum physics:

> If introspection is to be trusted, and it seems part of our very concept of mental states that it is trustworthy at least to this extent, then we are never in such superpositions.[28]

This is used to justify a riff on the many worlds interpretation called many minds. When I see the words "if introspection is to be trusted" I reach for my gun.

Another example comes from an article on determinism which begins "First, freedom … has to be asserted."[29] This is like starting an essay on planetary science by asserting the evident flatness of the earth. Then there are appeals to our perception, this is one of Wilson's justifications for the autonomy of high-level phenomena:

> Though the macro-entities of our acquaintance are, scientists tell us, materially constituted by massively complex and constantly changing microconfigurations, macro-entities do not perceptually appear to us as massively complex, constantly changing, configurations of microentities.[30]

Or appeals to the special sciences, here is philosopher David Robb:

> One argument for emergent mental causation starts with the frequently cited claim that psychology is an autonomous discipline.[31]

As if we could learn anything about the world from the limitations of our perceptual system or the existence of a discipline called psychology which (arguably) dates from the end of the nineteenth century and is likely one day to be swallowed by cognitive neuroscience.

---

[28] Albert and Loewer (1988).
[29] Nickel (2002).
[30] Wilson (2021), p. 5.
[31] Robb (2019).

Philip Clayton has written extensively on the significance of emergence for religion:

> Strong emergence—that is, emergence with downward causation—has the merit of preserving commonsense intuitions and corresponding to our everyday experience as agents in the world… for those who are idealists of a variety of stripes, and for theists who maintain that God as a spiritual being exercises some causal influence in the natural world, defending strong emergence may be a sine qua non for their position.[32]

Once more, subjective beliefs are being projected onto the world. More curious still:

> …depression, somatization, and the like, provide evidence that explanations for mood disorders and human behavior require irreducibly social and mental conditions and processes. Add to this the growing literature on placebo effects and conversion disorders… and we have plenty of prima facie evidence against [the causal closure of physics][33]

Let me end with an example from a physicist. Among the reasons that George Ellis gives that emergence must be real is "…because of the causal power of thoughts."[34] For the moment, I'll leave you to ponder this statement. I will return to it when I discuss mental causation in Chap. 15.

4.6 Further Reading

On fast and frugal heuristics and their adaptive nature, see Gigerenzer (2004). A recent review of intuitive physics is Kubricht et al. (2017) and an interesting discussion of its development through childhood can be found in Krist (2000). For the history of impetus, see Galili (2022), Chap. 2.

For a thorough discussion of the cognitive basis of science, see Carruthers (2006), Chap. 6. Wolpert (1994) is a thorough and readable account of how intuitions clash with science. An excellent popular book on cultural evolution is Henrich (2016). Kriegel (2003) is a review of the New Mysterianism and a discussion of the distinction between imaginative and representational understanding is in Vlerick and Boudry (2017). Ladyman and Ross

---

[32] Clayton (2006).
[33] Bishop et al. (2022), p. 295.
[34] Ellis (2016), p. 424.

(2009) discusses the mind projection fallacy in the context of a naturalistic metaphysics.

More suggestions for reading can be found at www.TheMaterialWorld.net.

References

Albert D, Loewer B (1988) Interpreting the many worlds interpretation. Synthese 77:195–213. https://doi.org/10.1007/BF00869434

Bauml B, Birbach O, Wimbock T, et al (2011) Catching flying balls with a mobile humanoid: System overview and design considerations. In: 2011 11th IEEE-RAS International Conference on Humanoid Robots. IEEE, Bled, Slovenia, pp 513–520

Bishop RC, Silberstein M, Pexton M (2022) Emergence in context: a treatise in twenty-first century natural philosophy. Oxford University Press, Oxford, United Kingdom

Boudry M, Vlerick M, Edis T (2020) The end of science? On human cognitive limitations and how to overcome them. Biol Philos 35:18. https://doi.org/10.1007/s10539-020-9734-7

Brault Foisy L-M, Potvin P, Riopel M, Masson S (2015) Is inhibition involved in overcoming a common physics misconception in mechanics? Trends in Neuroscience and Education 4:26–36. https://doi.org/10.1016/j.tine.2015.03.001

Carruthers P (2006) The architecture of the mind: massive modularity and the flexibility of thought. Clarendon Press; Oxford University Press, Oxford: Oxford; New York

Clark A, Chalmers DJ (1998) The Extended Mind. Analysis 58:7–19. https://doi.org/10.1093/analys/58.1.7

Clayton P (2006) Conceptual foundations of emergence theory. In: Clayton P, Davies PCW (eds) The re-emergence of emergence: the emergentist hypothesis from science to religion. Oxford University Press, Oxford; New York

Dennett DC (2017) From bacteria to Bach and back: the evolution of minds. W.W. Norton & Company, New York London

Ellis G (2016) How can physics underlie the mind? Springer, Berlin

Falcón-Cardona JG, Leguizamón G, Coello Coello CA, Castillo Tapia MaG (2022) Multi-objective Ant Colony Optimization: An Updated Review of Approaches and Applications. In: Dehuri S, Chen Y-W (eds) Advances in Machine Learning for Big Data Analysis. Springer Nature, Singapore, pp 1–32

Feynman RP (2006) QED: the strange theory of light and matter. Princeton University Press, Princeton, NJ

Galili I (2022) Scientific Knowledge As a Culture: The Pleasure of Understanding. Springer International Publishing AG, Cham

Gigerenzer G (2004) Fast and Frugal Heuristics: The Tools of Bounded Rationality. In: Koehler DJ, Harvey N (eds) Blackwell Handbook of Judgment and Decision Making. Wiley, pp 62–88

Griffiths TL (2020) Understanding Human Intelligence through Human Limitations. Trends in Cognitive Sciences 24:873–883. https://doi.org/10.1016/j.tics.2020.09.001

Hayek FA von, Bartley WW (2000) The collected works of F. A. Hayek. 1: The fatal conceit: the errors of socialism, ed. by W. W. Bartley III. University of Chicago Press, Chicago

Henrich JP (2016) The secret of our success: how culture is driving human evolution, domesticating our species, and making us smarter. Princeton University Press, Princeton

Jaynes ET (2010) Clearing up Mysteries — The Original Goal. In: Maximum Entropy and Bayesian Methods: Cambridge, England, 1988. Springer Netherlands, Dordrecht

Joyce J (2001) A portrait of the artist as a young man. Oxford University Press, Oxford

Kahneman D (2012) Thinking, fast and slow. Penguin Books, London

Kihlstrom JF (2004) Is there a "People are Stupid" school in social psychology? Behav Brain Sci 27:348–348. https://doi.org/10.1017/S0140525X04420081

Kriegel U (2003) The new mysterianism and the thesis of cognitive closure. Acta Anal 18:177–191. https://doi.org/10.1007/s12136-003-1020-1

Krist H (2000) Development of naive beliefs about moving objects. Cognitive Development 15:281–308. https://doi.org/10.1016/S0885-2014(00)00029-0

Kubricht JR, Holyoak KJ, Lu H (2017) Intuitive Physics: Current Research and Controversies. Trends in Cognitive Sciences 21:749–759. https://doi.org/10.1016/j.tics.2017.06.002

Ladyman J, Ross D (2009) Every thing must go: metaphysics naturalized. Oxford University Press, Oxford

Longino HE (1990) Science as social knowledge: values and objectivity in scientific inquiry. Princeton University Press, Princeton, N.

McCloskey M, Washburn A, Felch L (1983) Intuitive physics: The straight-down belief and its origin. Journal of Experimental Psychology: Learning, Memory, and Cognition 9:636–649. https://doi.org/10.1037/0278-7393.9.4.636

McLeod P, Reed N, Dienes Z (2003) How fielders arrive in time to catch the ball. Nature 426:244–245. https://doi.org/10.1038/426244a

Metzinger T (2009) The ego tunnel: the science of the mind and the myth of the self. Basic Books, New York

Mithen S (2002) Human evolution and the cognitive basis of science. In: Carruthers P, Stich SP, Siegal M (eds) The cognitive basis of science. Cambridge University Press, Cambridge ; New York

Nickel G (2002) Perspectives on Scientific Determinism. In: Atmanspacher H, Bishop RC (eds) Between chance and choice: interdisciplinary perspectives on determinism. Imprint Academic, Thorverton Exeter

Pinker S (1999) How the mind works. Norton, New York

Plato, Lee HDP, Lane MS (2007) The Republic. Penguin, London

Robb D (2019) Emergent mental causation. In: Gibb SC (ed) The Routledge handbook of emergence. Routledge, New York

Shaffer DM, McBeath MK (2002) Baseball outfielders maintain a linear optical trajectory when tracking uncatchable fly balls. Journal of Experimental Psychology: Human Perception and Performance 28:335–348. https://doi.org/10.1037/0096-1523.28.2.335

Sun Y (2019) Physarum-inspired Network Optimization: A Review. https://doi.org/10.48550/arXiv.1712.02910

Vlerick M, Boudry M (2017) Psychological Closure Does Not Entail Cognitive Closure: Psychological Closure Does Not Entail Cognitive Closure. Dialectica 71:101–115. https://doi.org/10.1111/1746-8361.12176

Wilson JM (2021) Metaphysical Emergence. Oxford University Press, Oxford

Wolpert L (1994) The unnatural nature of science. Harvard University Press, Cambridge, Mass

5

Computation and Simulation

Summary Computation and simulation are key parts of the scientific project and are also central to understanding emergence. After a general discussion of the role of simulations in science, the chapter turns to the theory of computation, Turing machines and the Church-Turing principle. Limits to the scope of simulation would imply limits to science. Absolute limits would arise if some things are non-computable. Thinking about practical limits leads us to quantum computers, quantum simulations and an estimation of the computational capacity of the universe. The physical version of the Church-Turing principle states that if we can build a universal quantum computer we can simulate every physical system.

Computation and its theory go deep into physics and particularly deep into the study of emergence. But the relation is not only theoretical. The last chapter described modern science as a system of distributed cognition, consisting of networked human minds using a whole range of cognitive scaffolds. Computers are one of these scaffolds and increasingly elements of the network in their own right.

Writing down models is a key part of the scientific project. Simulations are a way of investigating the properties of these models and will come up repeatedly in the following chapters. When I discuss the relation between emergence and models in Chap. 10 or the examples in Chap. 15, simulations are everywhere. Indeed, much of our theoretical knowledge of the world comes, in one way or another, from simulations.

© The Author(s), under exclusive license to Springer Nature
Switzerland AG 2025
L. Graham, *Physics Fixes All the Facts*, The Frontiers Collection,
https://doi.org/10.1007/978-3-031-69288-8_5

Limits to physics would imply limits to physicalism. If there are things beyond our understanding, then there is space for dualism. In Chap. 3, we saw limits involving the incompleteness of the scientific project. Then Chap. 4 asked whether we are cognitively closed to some things so will never be able to understand them. Given the centrality of simulations to science, a different way to address the question is to ask whether there are limits to simulations, whether there are systems we will never by able to simulate. Such limits may be absolute and understanding them requires understanding non-computability. Or they may be functions of the availability of computing resources and this leads to a discussion of quantum computation and quantum simulations. The central result is the Church-Turing-Deutsch principle which states that if we can build a universal quantum computer we can simulate every physical system. Let's start by discussing the nature of modelling and the role of simulations.

5.1   Wrestling with Models

The sciences do not try to explain, they hardly even try to interpret, they mainly make models[1]

Physics is about identifying laws, expressing them mathematically, then investigating their consequences.[2] Before the arrival of computers, there were essentially three ways of proceeding.

The first is to specify the model's equations and obtain a *closed-form* solution, one you can write down using a combination of well-known mathematical functions. Once you've got this, you can obtain numerical results to whatever precision you have time for. A standard example is the pendulum. A few lines of algebra give a closed-form expression which describes its motion.[3]

---

[1] Von Neumann (1955), p. 492.

[2] The question of why there are laws and why they can be expressed mathematically is a fascinating one, and I'll discuss it in Sect. 6.5.

[3] The expression is $\theta = \theta_0 \cos\left(\sqrt{\frac{g}{l}}t\right)$ where l is the length of the pendulum, g the gravitational acceleration, θ_0 the initial angle and θ the angle at time t. For more details see any classical mechanics textbook.

The second applies to systems which have large numbers of components. Then statistical methods can be used to average over the components to obtain aggregate quantities. Temperature and pressure are such aggregates. The accuracy of such averaging is related to the size of the system. In a litre of air, there are around 10^{22} molecules and the statistical calculation of its temperature will be highly accurate.

If neither of these are possible, you can try to build a physical model. J.D. Bernal gives this account from the late 1950s of his attempts to study liquids:

> I began, rather naively, by attempting to build models just to see what a structure satisfying these conditions would look like. I took a number of rubber balls and stuck them together with rods of a selection of different relative lengths ranging from 2.75 to 4 in. I tried to do this in the first place as casually as possible, working in my own office, being interrupted every five minutes or so and not remembering what I had done before the interruption.[4]

Bernal's model is a simulation, albeit a physical one. Just over a decade before he was struggling with balls and rods, the first general-purpose electronic computer, the Electronic Numerical Integrator and Computer or ENIAC had been built. Here's a rather breathless extract from the New York Times announcing it to the world in February 1946.

> In a matter of seconds, it does what trained computers [i.e. people] hitherto have taken weeks to perform... so clever is the device that its creators have given up trying to find problems so long they cannot be solved[5]

ENIAC was first put to work in the Manhattan Project, to simulate nuclear fission reactions. To understand whether a fissile material will reach criticality and undergo an explosive chain reaction, you need to track the paths of neutrons moving through it. The model consists of equations describing a neutron's motion and how it behaves if it collides with other particles. Solving the model would involve taking a large number of neutrons with a particular distribution of positions and energies, following them through a large number of collisions and calculating their resulting distribution. There is no closed-form solution and no satisfactory approximation.

---

[4] Bernal (1964). The results obtained using the model are reported in Bernal (1959).
[5] Kennedy (1946).

In March 1947, John von Neumann wrote a letter to a collaborator laying out the basic idea of a simulation. ENIAC could use the equations of the model to calculate the path of a small number of neutrons. Repeating this process many times for different starting values would approximate the statistical distribution implied by the model.

> Assume that one criticality problem requires following 100 primary neutrons through 100 collisions (of the primary neutron or its descendants) per primary neutron. Then solving one criticality problem should take about 5 hours. It may be, however, that these figures (100 × 100) are unnecessarily high. A statistical study of the first solutions obtained will clear this up. If they can be lowered, the time will be shortened proportionately.[6]

This extract captures the basic decisions and trade-offs involved in any simulation. The more particles and periods, the more accurate the results but the more computer time is needed. It's a matter of trial and error to find out what values you need to get a good enough simulation.

Despite some initial scepticism, computer simulations soon became common across the sciences. They can be found at every scale. Figure 5.1 shows graphical representations of the results of four types of simulation, chosen for being both recent and aesthetically pleasing. The left panel shows a simulation of the structure of a protein with the arrows showing its predicted motion. The middle panel shows a simulation of a whole cell. The red, white, and blue points represent strands of DNA, wrapped round the yellow spheres representing ribosomes and enclosed in the green cubes of the membrane. The right panel is a simulation of the early universe and galaxy formation, showing dark matter on the left transitioning to gas on the right. The bottom panel shows a molecular-level simulation of a fibre in a gecko's foot detaching from a surface.

---

[6] Quoted in Hurd (1985).

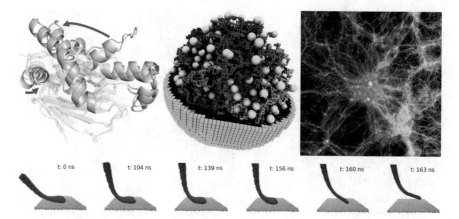

Fig. 5.1 Simulations[7]

While there are many formal techniques, both for building simulations and validating them, scientists are rarely purists and tend to make whatever assumption are necessary to get a model which serves their purposes. Many simulations are multi-scale, stitching together different types of models at different levels. More on this in Chap. 10.

There are two kinds of limits to what can be simulated. The first is theoretical and to understand this the next section introduces the theory of computation. The second relates to the availability of computational resources which I'll discuss later in the chapter.

Box 5.1 Algorithms and complexity

An algorithm is a systematic procedure to perform an operation. An example is the way addition is often taught. Write down the smaller number underneath the larger. Add the right-most digits. This gives the right-most digit of the answer. If it is greater than 10, carry 1. Repeat this for the second right-most digit. And so on.

In contrast, a heuristic is a rule of thumb that exploits shortcuts, demands less processing capacity but may be less reliable. When I add large numbers, I start with the leftmost digit and work to the right, stopping when the answer is as accurate as I need. This takes into account the limitations of my memory. If I lose track of where I am in the calculation, at least I'll be left with an estimate of the total correct to however many digits I managed. Famously,

---

the AI in Arthur C Clarke's 2001 was called HAL, Heuristically Programmed Algorithmic Computer, with the best of both worlds.

The idea of an algorithm underlies a useful way of characterising complexity. The Kolmogorov complexity of a system is defined as the shortest algorithm that produces the system as an output. So a 1000-digit number consisting only of the digit 1 can be compressed to the rule "Write '1' 1000 times", around 20 characters long. π can be calculated using the formula:

$$\frac{\pi}{4} = 1 - \frac{1}{3} + \frac{1}{5} - \frac{1}{7} + \ldots = \sum_{j=0}^{\infty} \frac{(-1)^j}{2j+1}$$

If you wrote this in words, you'd end up with around 60 characters.[8] So the infinite digits of π are highly compressible.[9] By contrast, a random number is by definition incompressible. The shortest algorithm that can produce a 1,000 digit random number x must be at least 1,000 characters long, e.g. "Write x". To see this, remember that random means that each digit is independent of all the others. If there were a shorter algorithm, this could be used to predict the next digit and so the number wouldn't be random.

The concept of compressibility will prove useful when discussing emergence since it can express the relation between the low level process which generates a system and its high level appearance. We'll see in the next chapter that the existence of laws of nature imply the universe is highly compressible.

5.2   Computability

Are there theoretical limits to simulation? If we restrict ourselves to simulations carried out on digital computers, this is asking if there are limits to what can be computed. The standard definition of computability was proposed in two papers published in 1936, one by Alonzo Church and the other by Alan Turing. The Church–Turing principle defines computability as follows:

Every 'function which would naturally be regarded as computable' can be computed by the universal Turing machine[10]

---

[8] For example "Calculate then write the sum over i from 0 to ∞ of 1 over 2 i + 1". Of course, for a formal definition of compressibility you need to specify the language of the algorithm. This will usually be a programming language.

[9] There is another mind-boggling algorithm that allows you to calculate any digit of π *without calculating the preceding ones*. See Bailey et al. (1997).

[10] The original thesis is in Turing (1937). This formulation is taken from Deutsch (1985).

The term "naturally" seems out of place in such a definition. But Turing's project was to define a machine which could replicate a human carrying out calculations which, as in the excerpt from the New York Times describing ENIAC, was the original sense of the term "computer". The human computer has pencil and paper and a state of mind. Together, the state of mind and the symbols on the paper determine what the person will do next.

Let's take an example. Imagine that you have in front of you a sheet of paper with a line of 1s and 0s and you are asked to decide if there is an odd number of 1s. If there are too many to do this at a glance, you may proceed by starting at one end of the string, when you meet the first 1 keeping the word "odd" in your head then changing it to "even" when you meet the next 1, and so on. I'll call this algorithm a parity checker.

Now let's look at how Turing conceived of a machine to carry out this algorithm. A Turing machine has several components. The first is a tape, divided into squares, each of which can contain a symbol. In Turing's paper the symbols can be anything, but let's restrict them to 0, 1, B for blank and H for halt. The second is a head which can move along the tape and both read from it and write to it. The third is an internal state, represented by a combination of the four symbols. The fourth is a set of rules which describes how the input changes the internal state and the output. This is shown schematically in the left panel of Fig. 5.2. In the right panel is a working implementation.[11]

For the parity checker, a Turing machine will use the rules shown in Table 5.1. The machine has two states, 0 and 1. These rules mean that if the machine is in state 0 it will remain in this state (row 1 in the table) until it comes across a 1 on the tape when it changes the state to 1 (row 2). If it is in state 1, it stays in the state (row 5) until it comes to a 1 when it changes

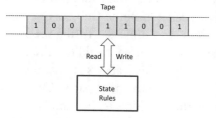

Fig. 5.2 A Turing machine[12]

---

[11] For an enlightening video of a Turing machine at work, see www.TheMaterialWorld.net.

[12] Right panel: Source: Rocky Acosta, own work, https://www.wikiwand.com/simple/Turing_machine#Media/File:Turing_Machine_Model_Davey_2012.jpg License: CC BY 3.0.

Table 5.1 A parity checker

State	Symbol on tape	New state	Write symbol	Move tape
0	0	0	–	Left
0	1	1	–	Left
0	B	H	0	–
1	0	1	–	Left
1	1	0	–	Left
1	B	H	1	–

the state to 0 (row 4). When it reaches a blank square, it writes the state to the tape and then stops (rows 3 and 6).

Turing's insight was that, whatever the computation, it could be formulated in this way, as a table with 5 columns or quintuples and a finite number of rows.[13] What makes a Turing machine so general is that it can be seen as a mathematical function, a procedure which, given an input, produces an output and whenever it is given the same input it produces the same output. Then a function is computable if we can build a Turing machine that, if it is given a tape containing only the input to the function, performs calculations then stops having written the output of the function to the tape.

5.3   Universal Turing Machines

Turing machines can compute a single function. A universal Turing machine is a Turing machine which can replicate any other Turing machine. When I described the parity-checking Turing machine, perhaps you took pencil and paper and worked through an example. You interpreted the table and this allowed you to carry out a simulation of the machine. You could just as easily have simulated any other description of a Turing machine (subject of course to time, memory and other constraints). You are a universal Turing machine.

To make a Turing machine universal, you just need to follow the same procedure. First, give it a set of internal rules which allows it to interpret sequences of quintuples. Then divide the tape into three parts. One part contains the set of quintuples describing the machine to be simulated, as in Table 5.1. The other two parts contain the result of the simulation, one to keep track of the state of the simulated machine and the other to write the tape of the simulated machine. Then let it carry out the simulation step by

---

[13] For more examples, see Minsky (1967), p. 120.

step, at each step writing the state of the simulated machine and its output onto the tape.

This is shown in Fig. 5.3. In modern computing terms, the rightmost part of the tape is the program, the internal rules of the machine are the interpreter and the rest of the tape is memory which holds the result of the simulation. From nothing but a clunky machine which can manipulate a paper tape, we've built a universal computing device.

> [Turing's] paper is significant ... because it contains, in essence, the invention of the modern computer and some of the programming techniques that accompanied it[14]

Babbage's analytical engine, powered by steam, would have been a universal Turing machine, had it ever been built. ENIAC was the first working universal Turing machine. The laptop I'm writing on is a universal Turing machine, as are most computational devices.[15] In a world of smartphones and high level languages, there is something charming about discussing a machine which uses paper tape. But the formulation is powerful precisely because it is so simple. The Church-Turing principle tells us that studying such machines is the same as studying computers in general. We can abstract away from the complex detail of computing. All computers do, down at the bottom, is shuffle the position of 0 and 1s.

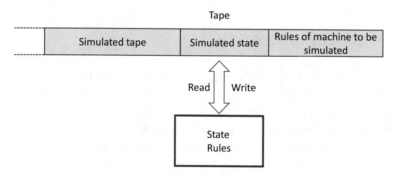

Fig. 5.3 A universal Turing machine

---

[14] Minsky (1967), p. 104.

[15] This is not strictly true since the theory of Turing machines assumes they can manipulate an unlimited amount of data whereas the memory of actual computers is limited. This has no practical import as one can always keep adding memory.

Box 5.2 Intractability

Imagine you are a delivery driver given a list of drop off points and the distances between them. How would you find the shortest route that visits every point then returns to the depot? This is the famous travelling salesperson problem.

The brute-force way to solve it is to try all possible combinations. If you have 4 points as well as the depot, when you start you have a choice of 4, from the next one a choice of 3, from the next a choice of 2. So there are 4 × 3 × 2 = 4! = 24 combinations, or 12 unique combinations since you can go round each route in two directions. The number of combinations explodes with the number of points. For 30 points, not a huge delivery run, there are around 10^{32} combinations. If you calculated one every nanosecond, the time since the big bang wouldn't be enough to run through all of them. Such problems are known as NP-hard and set a limit on what can be computed.[16] Future increases in (classical) computing power won't help you since real world problems with tens of thousands of nodes would requires more computing resources than exist in the whole universe.

Such intractability constrains simulations less than it might seem. There are often shortcuts or heuristics that can find solutions or approximations. For example, in 2018 a group found the shortest route between the nearly 50,000 pubs in the United Kingdom.[17] Also, when it comes to simulating physical systems, it's important to remember that nature may be subject to the same computational restrictions. This means that if we find ourselves trying to simulate an NP-hard problem, we probably haven't found the smartest way. I'll return to this in the discussion of strong emergence in Chap. 11.

5.4   Non-computability

Are there things which can't be computed? Box 5.2 shows that some problems might be intractable due to lack of computing power. This section asks a stronger question: are there any absolute limits to the problems that can be solved by a Turing machine and hence by any digital computer?

Turing's original paper included several examples of non-computable functions. In the 1950s these were reformulated as the *halting problem*. Any given Turing machine, or if you like, any program, will either produce a result then stop or continue running for ever. Is there a general algorithm to tell if a particular Turing machine will stop or not? Turing proved that no such method can exist. So the halting problem is *non-computable* or *undecidable*.

---

[16] NP stands for "Nondeterministic Polynomial-time" linked to a class of computing devices called non-deterministic Turing machines. See Harel and Feldman (2004), Chap. 7.

[17] https://www.math.uwaterloo.ca/tsp/uk/index.html.

Here is a simple proof by contradiction. Say you have a program that claims it solves the halting problem, let's call it HALT. As input, it takes the program to be tested, P, and returns 1 if the program halts and 0 if not. Then define the program P as IF HALT(P) = 1 THEN NEVER STOP. However sophisticated the program HALT, our one line program will always defeat it. So we have proved that no general solution to the halting problem is possible. This might be called proof by malicious program.[18]

Box 5.3 A halting problem

Here is a computable algorithm.

1. Start with a positive integer, X
2. If X = 1 stop
3. X = X − 2
4. Go to step 2

It is easy to see this will stop only if the initial value of X is odd. For even numbers, at some point you will reach 2, then 0 and then on forever without ever satisfying the X = 1 condition which makes the algorithm stop.
  Now let's modify the algorithm:

1. Start with a positive integer, X
2. If X = 1 stop
3. If X is even, X = X / 2
4. If X is odd, X = 3X + 1
5. Go to step 2.

This is called the Collanz conjecture. For even numbers, the algorithm clearly stops. For odd numbers, it produces long sequences. For example, starting with 11 gives 34, 17, 52, 26, 13, 40, 20, 10, 5, 16, 8, 4, 2, 1. If you start at 27, the sequence continues for 111 steps. No one has ever found a number for which the system doesn't stop. Yet neither has anyone found a proof that there is no such number. In the absence of such a proof, the only way to find out whether the algorithm stops for a particular number is to run the algorithm. And, however many iterations you run it for, you can never know whether just one more will start it on the path to stopping. This is the essence of the halting problem.

More generally, it is easy to see that there are infinitely many non-computable functions. Every Turing machine can be specified by an integer.

---

[18] The proof is due to Strachey (1965). Its self-referential nature may make you think of Gödel's incompleteness theorem. There is indeed a deep link with non-computability, for a discussion, see Penrose (1999), Chap. 2 and also Box 5.4.

To see this, look back at Table 5.1. You can write each row as a line of 0 and 1 s. To incorporate H and B, you'll need a two-bit language in which 0 is 00, 1 is 01, B is 10 and H is 11. That makes 10 bits for each row. The table has 6 rows, so this Turing machine can be described by a binary number 60 bits long.[19] This integer can be thought of as the program you'd need to give a universal Turing machine so it can simulate our parity checker.

This applies to any computing device. Take the computer in front of you. Its capacities are fully defined by the contents of its memory and the programs hard coded into its processors, all of which can be represented by strings of 0 s and 1 s. Write down every bit in the memory and the processors as a long string and you have a (very large) integer which specifies your computer. As I type this sentence, the contents of my laptop's memory change and so does the integer describe the computer. Switch from word processor to browser and a part of the sequence of 1 and 0 s changes and so does the integer. At any moment your computer or any Turing machine can be fully described by an integer.

We saw that each Turing machine represents a mathematical function. But mathematical functions can be expressed as real numbers. Since there are infinitely many more real numbers than integers, there are more functions than there are Turing machines.[20] This means an infinite number of mathematical functions must be non-computable since they do not correspond to a Turing machine. This might seem abstract. Indeed a list of non-computable problems is intimidating to a non-mathematician.[21] But the idea is profound. However smart you are, however much computing power you have at your disposal, there are well defined problems which cannot be solved,

If we turn to physics, an immediate implication of non-computability is that Turing machines, and hence computers in general, cannot perfectly simulate classical physics. Classical quantities are continuous, taken from the real numbers, so an infinite number of them will not correspond to Turing machines described by an integer.[22]

---

[19] If you really want to know, it's 011011010001010001000100010000001011000000101000000 00000000 or in decimal 493,500,697,503,941,000.

[20] The proof that there are more mathematical functions than there are integers and that there are more real numbers than integers is essentially the same. It is due to Georg Cantor and uses the elegant diagonalization method. For a clear explanation see Gardner (2001), p. 331.

[21] An interesting exception is the impossibility of determining the algorithmic complexity (Box 5.1) of a general string, the proof of which is closely linked to the halting problem.

[22] On the other hand, an analogue computer could in principle simulate any classical system. This shows the link between a particular concept of computability and a particular type of computer. I've been using computation in Turing's sense, as something that can be computed by a Turing machine. But analogue computers cannot in general be represented by Turing machines so imply a different sense of computability.

However we live in a non-classical world. In such a world, is anything that is of interest to a physicist rather than a mathematician non-computable? I'll return to this question in Chap. 11 when I discuss whether non-computability can be a form of strong emergence. Always lurking behind the idea of non-computability is the difference between the infinity of the real numbers and the infinity of the integers. This means that everything in a finite system is computable. Even in an infinite system in which some functions are non-computable, you can always find a computable function that approximates them arbitrarily well in the same way there is always a rational number arbitrarily close to a real number. So it's hard to see the issue having any importance for questions of simulation.

Box 5.4 Is the brain a hypercomputer?

A system that can solve a non-computable problem is known as a *hypercomputer*. A discussion can be found in Turing's work but I think the claim that the brain is a hypercomputer was first stated explicitly by J.R. Lucas[23] in 1961. Subsequently Roger Penrose has made it a central part of his quantum theory of consciousness. Loosely, the argument goes as follows:

Take a proposition P which states that there is no proof of proposition P.

- P cannot be proved false, since, in Penrose's words "Our formal system should not be so badly constructed that it actually allows false propositions to be proved!"[24]
- P cannot be proved true, since this would be contradictory
- Thus neither P nor its opposite is provable. So P is non-computable.
- This establishes the truth of P

This is pretty neat. We've proved that something cannot be proved but in doing so we have showed it is true. We have proved something non-computable. We are hypercomputers!

There seems to be broad agreement that there is something wrong with this, but no consensus as to what. More importantly, such an argument, with its Alice in Wonderland aspect, seems a fragile hook on which to hang a staggeringly important idea. If the brain is a hypercomputer, and if like Penrose we are good physicists, this means we need to find some new physics to explain it. This has motivated much of Penrose's work on the relation between quantum physics and consciousness.

---

[23] Lucas (1961).
[24] Penrose (1999), p. 108.

5.5   Quantum Computation

So far, this has all been about classical computation. Information is represented by bits that can be either 0 or 1. Quantum computation uses qubits which can be in superpositions, states which are 0 and 1 simultaneously. To see the difference between the approaches, consider the following problem. You are inviting a bunch of friends to dinner and know which of them get on and which of them don't. The problem is to allocate them between the two tables in your living room to maximise the number of pairs who get on and minimise the number of those who don't.

Let's take the simplest example where you have 3 friends (A, B and C). You know than A and B get on; B and C clash and A and C clash. To rank outcomes, let's define a payoff which is the number of pairs which get on minus the number of pairs which clash. You can represent a seating plan by 3 bits, where e.g. 110 means A and B are at table 1 and C at table 0. Table 5.2 shows the 8 possible seating plans.

Although in this case the answer is obvious (put A and B at one table and C at the other), this will not be the case for larger numbers of people or tables. A classical computer algorithm to solve the problem in general will need to work through the possible combinations and calculate the payoff function each time. For N people, this will require the payoff function to be calculated 2^{N-1} times (it is $N - 1$ since the problem is symmetric, the bottom four lines of the table are the same as the top four so you only have to calculate half of the possibilities). For $N = 100$, this requires the payoff function to be calculated around 10^{30} times which puts it well beyond the limits of the fastest computer (which, at the time of writing, manages around 10^{18} operations per second).

Table 5.2   Three guests and two tables

[ABC]	Payoff
000	−1
001	1
010	−1
011	−1
100	−1
101	−1
110	1
111	−1

Instead, set up a quantum computer with 3 qubits in superposition i.e. they simultaneously take the 8 possible values. Then apply the payoff function so that when the superposition collapses it provides one of the solutions. If you increase the number of people, you need to increase the number of qubits but you still only need to calculate the payoff function once. In principle, a quantum computer should be able to solve the problem for a hundred or a million people at the same speed. This example brings out the way quantum computers can solve problems in parallel, performing a calculation on all possible states at once.

Of course, in practice things are more complicated than that but the details don't need to trouble us here. It's general accepted that quantum computation has staggering speed advantages for particular types of problem.

Are quantum computers subject to the same limitations of computability as classical computers? Yes. Just as in the case of a Turing machine, you can always use pencil and paper to replicate its workings, given enough time.[25] In other words, you can always program a classical computer to simulate a quantum computer. The simulation might be painfully inefficient, but that's not the point. If there is a Turing machine corresponding to every quantum computer, then the limits of computability are not changed.

5.6   Quantum Simulation

In 1982, Richard Feynman published one of the foundational papers of quantum computing, entitled"Simulating physics with computers". Here is its closing sentence:

> Nature isn't classical, dammit, and if you want to make a simulation of nature, you'd better make it quantum mechanical, and by golly it's a wonderful problem, because it doesn't look so easy.[26]

Given you have a quantum computer to hand, there are two general ways of using it as a simulator. The first involves encoding the system of interest as an array of qubits then finding a quantum algorithm which efficiently (and perhaps approximately) calculates its time evolution while benefitting from the speed increases due to the massively parallel nature of the calculation. Return to the example in the previous section. Instead of three people and two tables, take a quantum system of three particles each of which can take

---

[25] For a proof of this, see Nielsen and Chuang (2010), sec. 4.5.5.
[26] Feynman (1982).

two states. A classical computer would need to solve the relevant equations of quantum physics eight times whereas a quantum computer would need to solve them once. A system of 100 particles is far beyond the power of any plausible classical computer. A quantum computer could, in principle, solve it with a single calculation.

But Feynman was thinking of something different. Instead of using a quantum algorithm, a second technique involves configuring the quantum computer to exactly represent the system being simulated. You can then exploit the high level of control given by a quantum computer to study the system's evolution.

The first of these is sometimes called a *digital quantum simulation* the second is an *analogue quantum simulation*. Examples of both can be found in the recent literature. What's fascinating is this. In the second case, the distinction between simulation and system has all but vanished. The simulation has become an experiment. Far from the complications described by Bernal in his ball and rod model of liquids, a general-purpose quantum computer would give an exquisite level of control over the atomic details of what's going on.

5.7   The Church-Turing-Deutsch Principle

A Turing machine takes inputs and by a deterministic process gives outputs. A physical system can be described in the same way. The inputs are initial conditions, the function the structure of the system and the outputs the final state. For a classical example, take of a set of billiard balls on a table. The initial conditions are the starting position of the balls and the force and direction with which one is hit. The function consists of Newton's laws and a description of the properties of the table. The output is the final resting position of the balls. For a quantum system, and so for everything, the input is the initial quantum state, the function contains the laws of quantum physics and the output is the final measurement.

David Deutsch, in another foundational paper of quantum computation, combines this parallel between computation and physical systems with the Church-Turing principle to obtain what has come to be known as the Church-Turing-Deutsch principle:

> Every finitely realizable physical system can be perfectly simulated by a universal model computing machine operating by finite means[27]

---

[27] Deutsch (1985).

Underlying this is Deutsch's concept of a universal quantum computer. It can perfectly simulate any physical system using an algorithm which makes it indistinguishable from the system. This digital simulation can exactly reproduce any analogue simulation. He formally describes such a universal quantum computer and shows that its existence collapses the distinction between experimental physics and computer science. Since a program exists for every physical process, studying a system at an atomic resolution is a matter of finding the right program.

5.8   The Limits of Physics 3

The Church-Turing-Deutsch principle gives another way to think about the scope of physics. Deutsch says that it means that if humans can build universal quantum computers they can become universal explainers. If every physical system can be simulated, every physical system can be explained. To assert that there are inexplicable phenomena is to assert that there are systems which are supernatural and not subject to the laws of physics.

This is an extremely strong contention. Is simulating a system the same as understanding it? In an interesting passage, Anderson writes that in 1956 he heard Richard Feynman say he had no idea what caused superconductivity:

> ... it was not lack of computational power in any real sense that was the obstacle. If we had actually had the (actually impossible) ability to follow the motions of all of the electrons and ions in detail, all that the computer output could have told us would have been that the material was exhibiting all of the well-known and well-studied phenomenology of superconductivity: it could not have told us why, because it would not know what that question means. What it does mean is that there are certain concepts and constructs which allow enormous compression of the brute-force calculational algorithm, down to a set of ideas which the human mind can grasp as a whole.[28]

I take the point to be that even if we can simulate something, we won't necessarily have imaginative understanding of it. This requires simplified models. Would we have representational understanding? The passage shows that Anderson is thinking of a classical simulation made up of a series of equations which are solved or approximated. Instead, a quantum simulation would be like an experiment over which we have precise control of

---

[28] Anderson (2011), p. 136.

every detail. Does this necessarily imply it would give us representational understanding? I'll return to this question in Chap. 14.

But this aside, when we think of simulations, we don't have to worry about non-computability because we are only concerned with finite systems. Neither do we need to worry about intractability since these limits also apply to the physical systems we are simulating. In practice, of course, computation is limited by constraints of energy, time and technological capacity. The next section discusses what physics can tell us about these.

5.9   Ultimate Limits

In a 1996 paper[29] Rolf Landauer showed that information processing must dissipate energy. Information is not abstract but is always instantiated in some physical system. This could be an arrangement of beads on an abacus, pencil marks on a sheet of paper or patterns of magnetism on the hard drive of my laptop. Information processing involves changing information and hence the physical system in which it is stored. Like any physical process, this must dissipate energy. This has a whole host of interesting implications, but here what matters is that it allows the calculation of upper bounds on computational capacity of physical systems.

Putting aside non-computability, the scope of simulations is limited by the computational resources available. In a pair of papers,[30] Seth Lloyd estimates the computational capacities of a couple of systems. The first system is what he calls the ultimate laptop consisting of 1kg of matter. He uses the Uncertainty Principle to derive an expression for the minimum energy needed for a qubit to transition from one state to another in a given time. We can calculate the total energy of the ultimate laptop from $E = mc^2$ and use this to obtain the maximum number of operations it can perform. This turns out to be around 10^{50} per second with a maximum memory space of 10^{31} bits.

In a second paper, Lloyd repeats the calculation for the universe as a whole to estimate an upper bound to the amount of computation carried out since the big bang. The answer is around 10^{120} operations. This is an interesting number since it is also a lower bound on the number of operations a quantum computer would need to perform to perfectly simulate the universe. Does it represent the number of calculations that the universe has actually performed? Does the universe perform calculations at all? I turn to this question in the next section.

---

[29] Landauer (1996).
[30] Lloyd (2000) and Lloyd (2002).

Box 5.5 Quantum computing and the multiverse

While we are far from being able to build large quantum computers, there seems little doubt that the underlying theory is correct. A standard example of the superiority of quantum computers is their ability to use a technique called Shor's algorithm to factorise large numbers so much faster than classical computers that they could potentially break many standard forms of encryption.

But computation must be realised in a physical system, using physical resources. Information must be stored and energy expended in processing it. For a classical computer, you can measure these resources by the amount of power it consumes. For a quantum computer, things are more complicated. Here's David Deutsch:

> I issue this challenge: explain how Shor's algorithm works. I do not merely mean predict that it will work, which is merely a matter of solving a few uncontroversial equations. I mean provide an explanation. When Shor's algorithm has factorized a number, using 10^{500} or so times the computational resources that can be seen to be present, where was the number factorized? There are only about 10^{80} atoms in the entire visible universe, an utterly miniscule number compared with 10^{500}. So if the visible universe were the extent of physical reality, physical reality would not even remotely contain the resources required to factorize such a large number. Who did factorize it, then? How, and where, was the computation performed?[31]

Deutsch's answer to this question is that the computation happens in the multiverse. Quantum computers are massively parallel, with each universe computing one possible outcome. The result is generated by a process of interference which fuses the different universes involved in the calculation and corresponds to the most probable outcome.

5.10   The Great Programmer

Lloyd's work shows that the universe is analogous to a quantum computer that at each moment calculates its own evolution.[32] The universe may be a computer, but what if it is also the result of a computation?

One of my favourite articles[33] on the subject of computability starts from the observation that any program can be represented by an integer. Simulations are no more than a particular sort of program, so they too can

---

[31] Deutsch (1998), p. 217.

[32] Lloyd (2013).

[33] Schmidhuber (1997).

be represented by an integer. If the Church-Turing-Deutsch principle is true and every physical system can be simulated, it can also be represented by a (very long) integer. Running the simulation is a matter of writing down this integer. This applies to our universe or to any possible universe.

The Great Programmer has a universal Turing machine. As inputs and outputs, the machine takes three symbols, 0, 1 and a comma. The comma represents the division between periods of time in the simulation. The possible input programs for the machine then consist of all the possible combinations of these three symbols and the output will be another combination of these symbols, which may be finite or infinite. Each of these outputs corresponds to a simulated universe.

The Great Programmer runs all possible programs on its Turing machine and so simulates all possible universes. What do these universes look like? Some will be finite (the program producing them stops at some point); some will be infinite. Most will be random sequences of the three characters resembling noise. Some universes will evolve according to rules and sometimes these rules will resemble physical laws.

This is just a restatement of the idea of algorithmic incompressibility. Random sequences are incompressible. But a universe governed by physical laws is compressible: its state is calculated by applying a program describing the laws to its previous state. However there is no guarantee that physicists within the universe if there happen to be any, will be able to find this program (the true description of the universe). They may identify as noise or indeterminism properties that from the perspective of the Great Programmer are law like:

> Our fundamental inability to perceive our universe's state does not imply its true randomness, though. For instance, there may be a very short algorithm computing the positions of electrons lightyears apart in a way that seems like noise to us but is in fact highly regular.[34]

The paper concludes by observing that most philosophical problems dissolve from the perspective of the Great Programmer. The many worlds interpretation of quantum mechanics comes for free. Is there life after death? In some universes, yes, in some no. Discussions about body, soul or consciousness, are, to the Great Programmer, nothing but particular strings that may arise in many universes. They may allow the inhabitants of the universes to discuss things that interest them, but to the Great Programmer they may be no more relevant than a string describing a star or some random

---

[34] Schmidhuber (1997).

noise. There is no reason why the Great Programmer should be concerned that strings representing life evolve in some universes.

The problem of time goes away too:

> The Great Programmer does not worry about computation time. Nobody presses Him (sic). Creatures which evolve in any of the universes don't have to worry either. They run on local time and have no idea of how many instructions it takes the Big Computer to compute one of their time steps, or how many instructions it spends on all the other creatures in parallel universes.[35]

One way to respond to this argument is that it has no empirical consequences. Another is to say that simulation is different from the real thing; simulating water on my computer can never make me wet. While that is surely the case, are we sure that if we added creatures to the simulation we could not simulate their feeling wet? But this brings us smack up against the "hard" problem of consciousness and so is a good place to end this discussion.

5.11 How Does the Universe Do It?

Pull the plug out of a bathtub, and you get a vortex. This involves solving a quantum many-body problem involving 10^{24} or so quarks and electrons. The universe certainly gives the impression of solving this problem quickly and reliably. The description of a universe as an analogue quantum computer would say that the 10^{24} fundamental particles have exactly the computing power to do this, but the question stands: how do they do it?

You can ask a similar question in terms of physics: how do particles "know" how to behave? Physics can answer this to a certain extent. One example would be the demonstration that energy-minimising trajectory of classical particles is a result of a quantum summation over all possible trajectories. Why do particles travel in straight lines? Because they try out all possible paths and the non-straight ones cancel each other out. But then we can ask why particles try out all possible paths and the question slips into metaphysics.

However, in the domain of computing the question seems well posed: how does the universe do it? The question stands whether you think the universe is a simulation (in which case it applies to the universe the simulation is running in) or whether you think the universe is a computation. We'll come across two possible answers as examples of strong emergence in Chap. 11, both of

---

[35] Schmidhuber (1997).

which contend that the universe mostly avoids doing quantum computations. Perhaps the answer lies with the multiverse (Box 5.5). But you can turn this round and argue that if the universe has an efficient method of computation, Deutsch's argument for the multiverse dissolves.

My judgement would be that as far as this question goes we are not much further forward than Empedocles was about gravity when he asserted that bodies fall to the earth out of love for their fellow material object.

5.12   Further Reading

For a thorough discussion of the role of models in science, see Gelfert (2016) and for philosophical perspectives on the role of simulations, Lenhard (2019) or Winsberg (2022). Borwein and Crandall (2013) is a lively discussion of the idea of closed-form solutions. For a more in depth discussion of the concept of an algorithm along with useful examples, see Penrose (1999), Chap. 2. A fascinating discussion of the wide-ranging philosophical implications of simulations is to be found in Chalmers (2022).

Harel and Feldman (2004) is an excellent introduction to many of the topics covered in this chapter, including Turing machines, the theory of computability and NP-hard problems. For more on the role of human computers in Turing's project, see Lupacchini (2018) and on the halting problem Lucas (2021). For non-technical discussions of Penrose's argument for the non-computability of the brain, see Chalmers (1995).

Introductions to quantum computation can be found in Lloyd (2007) and Deutsch (1998), Chap. 9. On quantum simulation, see Buluta and Nori (2009), Johnson et al. (2014) and Fauseweh (2024).

More suggestions for reading can be found at www.TheMaterialWorld.net.

References

Anderson PW (2011) More and different: notes from a thoughtful curmudgeon. World Scientific, New Jersey

Bailey D, Borwein P, Plouffe S (1997) On the rapid computation of various poly-logarithmic constants. Math Comp 66:903–913. https://doi.org/10.1090/S0025-5718-97-00856-9

Bernal JD (1959) A Geometrical Approach to the Structure Of Liquids. Nature 183:141–147. https://doi.org/10.1038/183141a0

Bernal JD (1964) The Bakerian Lecture, 1962 The structure of liquids. Proc R Soc Lond A 280:299–322. https://doi.org/10.1098/rspa.1964.0147

Borwein JM, Crandall RE (2013) Closed Forms: What They Are and Why We Care. Notices Amer Math Soc 60:50. https://doi.org/10.1090/noti936

Buluta I, Nori F (2009) Quantum Simulators. Science 326:108–111. https://doi.org/10.1126/science.1177838

Chalmers DJ (1995) Minds, Machines, And Mathematics: A Review of Shadows of the Mind by Roger Penrose. Psyche 2. https://journalpsyche.org/files/0xaa25.pdf

Chalmers DJ (2022) Reality+: virtual worlds and the problems of philosophy. W. W. Norton & Company, New York (N.Y.)

Deutsch D (1985) Quantum theory, the Church–Turing principle and the universal quantum computer. Proc R Soc Lond A 400:97–117. https://doi.org/10.1098/rspa.1985.0070

Deutsch D (1998) The fabric of reality: the science of parallel universes-- and its implications. Penguin Books, New York

Fauseweh B (2024) Quantum many-body simulations on digital quantum computers: State-of-the-art and future challenges. Nat Commun 15:2123. https://doi.org/10.1038/s41467-024-46402-9

Feynman RP (1982) Simulating physics with computers. Int J Theor Phys 21:467–488. https://doi.org/10.1007/BF02650179

Gardner M (2001) The colossal book of mathematics: classic puzzles, paradoxes, and problems: number theory, algebra, geometry, probability, topology, game theory, infinity, and other topics of recreational mathematics. Norton, New York

Gelfert A (2016) How to do science with models: a philosophical primer. Springer, Cham

Harel D, Feldman YA (2004) Algorithmics: the spirit of computing. Addison Wesley: Pearson Education, Harlow, Essex, England; New York

Hurd CC (1985) A Note on Early Monte Carlo Computations and Scientific Meetings. IEEE Annals Hist Comput 7:141–155. https://doi.org/10.1109/MAHC.1985.10019

Johnson TH, Clark SR, Jaksch D (2014) What is a quantum simulator? EPJ Quantum Technol 1:10. https://doi.org/10.1140/epjqt10

Kennedy, Jr. TR (1946) Electronic Computer Figures Like a Flash. New York Times.

Landauer R (1996) The physical nature of information. Physics Letters A 217:188–193. https://doi.org/10.1016/0375-9601(96)00453-7

Lenhard J (2019) Calculated surprises: a philosophy of computer simulation. Oxford University Press, New York, NY

Lloyd S (2000) Ultimate physical limits to computation. Nature 406:1047–1054. https://doi.org/10.1038/35023282

Lloyd S (2002) Computational Capacity of the Universe. Phys Rev Lett 88:237901. https://doi.org/10.1103/PhysRevLett.88.237901

Lloyd S (2007) Programming the universe: a quantum computer scientist takes on the cosmos. Vintage Books, New York , NY

Lloyd S (2013) The universe as quantum computer. https://doi.org/10.48550/ARXIV.1312.4455

Lu W, Zhang J, Huang W, et al (2024) DynamicBind: predicting ligand-specific protein-ligand complex structure with a deep equivariant generative model. Nat Commun 15:1071. https://doi.org/10.1038/s41467-024-45461-2

Lucas JR (1961) Minds, Machines and Gödel. Philosophy 36:112–127. https://doi.org/10.1017/S0031819100057983

Lucas S (2021) The origins of the halting problem. Journal of Logical and Algebraic Methods in Programming 121:100687. https://doi.org/10.1016/j.jlamp.2021.100687

Lupacchini R (2018) Church's Thesis, Turing's Limits, and Deutsch's Principle. In: Cuffaro ME, Fletcher SC (eds) Physical Perspectives on Computation, Computational Perspectives on Physics. Cambridge University Press, pp 60–80

Materzok T, De Boer D, Gorb S, Müller-Plathe F (2022) Gecko Adhesion on Flat and Rough Surfaces: Simulations with a Multi-Scale Molecular Model. Small 18:2201674. https://doi.org/10.1002/smll.202201674

Minsky M (1967) Computation: finite and infinite machines. Prentice-Hall, Englewood Cliffs, NJ

Nielsen MA, Chuang IL (2010) Quantum computation and quantum information. Cambridge University Press, Cambridge

Penrose R (1999) The emperor's new mind: concerning computers, minds and the laws of physics. Oxford University Press, Oxford

Schmidhuber J (1997) A computer scientist's view of life, the universe, and everything. In: Freksa C, Jantzen M, Valk R (eds) Foundations of Computer Science. Springer Berlin Heidelberg, Berlin, Heidelberg, pp 201–208

Strachey C (1965) An impossible program. The Computer Journal 7:313–313. https://doi.org/10.1093/comjnl/7.4.313

Thornburg ZR, Bianchi DM, Brier TA, et al (2022) Fundamental behaviors emerge from simulations of a living minimal cell. Cell 185:345-360.e28. https://doi.org/10.1016/j.cell.2021.12.025

Turing AM (1937) On Computable Numbers, with an Application to the Entscheidungsproblem. Proceedings of the London Mathematical Society s2–42:230–265. https://doi.org/10.1112/plms/s2-42.1.230

Von Neumann J (1955) Method in the Physical Sciences. In: Collected works. 6: Theory of games, astrophysics, hydrodynamics and meteorology. Pergamon Press, Oxford

Winsberg E (2022) Computer Simulations in Science. The Stanford Encyclopedia of Philosophy. https://plato.stanford.edu/entries/simulations-science/

Part II

The Illusion of Emergence

6

Weak Emergence: More is Different

Summary "More is different" has become a slogan for proponents of emergence and the idea is central to the way many condensed matter physicists think about their discipline. This chapter argues that it fails as a definition of emergence because it applies to any composite system. More is always different. Understanding how parts interact to produce the behaviour of wholes has always been at the heart of physics.

The attraction of weak emergence is that it promises a non-reductive physicalism. This would allow us to avoid both the supernatural and the violence that physicalism does to our commonsense conceptions of the world and ourselves. The five chapters of Part II discuss and criticise various concepts of weak emergence. This chapter deals with "more is different". The next chapter discusses simulation emergence, Chap. 8 turns to multiple realization emergence and Chap. 9 to contextual emergence. The examples used in these chapters are mostly drawn from physical systems. A discussion of the relation between weak emergence and the models that scientists use is left to Chap. 10.

Before starting, take a glance back at Sect. 3.13 where I showed how non-reductive physicalism can be rejected with a 4-line argument. If you find that argument convincing, you can safely skip to Chap. 11 and the discussion of strong emergence.

"More is different" was the working definition I proposed at the start of Chap. 2 and this makes the article that introduced it a good place to begin. I'll then turn to two related definitions. Despite the clarity and appeal of

© The Author(s), under exclusive license to Springer Nature
Switzerland AG 2025
L. Graham, *Physics Fixes All the Facts*, The Frontiers Collection,
https://doi.org/10.1007/978-3-031-69288-8_6

these definitions, it is easy to see that they apply to all composite systems. More is always different, and physics has always been about describing the interactions between parts which give wholes their distinct properties.

6.1 The Reemergence of Emergence

"More is different" is the title of an influential article published in 1972 by Philip Anderson (Nobel Prize for Physics, 1977). It reintroduced emergence into the mainstream and the phrase has become a slogan for proponents of emergence. It also helped constitute the discipline of condensed matter physics as the title of a book celebrating the paper's 50th anniversary make clear: "More is Different: Fifty Years of Condensed Matter Physics.[1]

Anderson's point is about the difference between reductionism and what he calls constructionism. The paper start with a rousing defence of reductionism then argues that while we can always work down from wholes to parts, we can't work up from part to wholes. The properties of the whole arise not just from the properties of the parts but from their mutual interaction:

> The behavior of large and complex aggregates of elementary particles, it turns out, is not to be understood in terms of a simple extrapolation of the properties of a few particles. Instead, at each level of complexity entirely new properties appear, and the understanding of the new behaviors requires research which I think is as fundamental in its nature as any other.[2]

The examples Anderson uses are either symmetry breaking as in Sect. 2.5 or the thermodynamic limit which I'll discuss in detail in Chap. 10. Since Anderson accepts reduction, the message of the paper is that construction may be beyond our computational or cognitive abilities.

6.2 Emergence in Condensed Matter Physics

The phrase "more is different" only appears in the title of the Anderson's article and the term emergence does not appear at all so let me give two related definitions. Between them, they put some structure on the idea that

---

[1] Ong and Bhatt (2001).
[2] Anderson (1972).

"more is different" and capture what I take to be the standard view of emergence among condensed matter physicists. The first, due to Navot Israeli and Nigel Goldenfeld, is

> Emergent properties are those which arise spontaneously from the collective dynamics of a large assemblage of interacting parts.[3]

The second is from Robert Laughlin who we've already come across in the context of the fractional quantum Hall effect. In an interview, he defines emergence as:

> A collective principle of organization that gives rise to a law, a relationship among measured things that is always true.[4]

More is different in the sense that when you have more, you have new empirical regularities, new laws. A quantitative difference becomes a qualitative difference. An interesting aspect, which Laughlin emphasises in the interview, is the link between emergence and experiment. If you measure some law-like relation at the level of the system as a whole, then this relation is emergent, the result of self-organization of the parts.

The link to experiment means the notion of what is fundamental goes away. Laws are empirical regularities and that is that. Laughlin asserts that there is no reason to prefer one level to another and attempts to do so are "ideology, the results of a belief system".[5]

Both these definitions are connected the idea of multiple realization which is the subject of Chap. 8. There is also a link to Chalmers's definition of emergent phenomena as "unexpected" (Sect. 3.12). Until we do the experiment, the properties of wholes are unexpected given what we know about the parts. We can look at the parts as much as we like, but until experiment teaches us about their complex mutual interactions the behavior of the whole will remain a mystery.

---

[3] Israeli and Goldenfeld (2006).
[4] Laughlin (2021).
[5] Laughlin (2021).

6.3   More is Always Different

Take a look back through the examples in Chap. 2. They all display behaviour distinct from that of their parts so match Anderson's sense of "more is different". They all involve empirical regularities at the level of the whole so fit Laughlin's definition. For all of them, mutual interactions between the components are key to understanding the properties of the system. By necessity, a composite system involves interactions between its parts and by necessity these interactions mean the parts behave differently than they would in isolation. It seems these definitions apply to all composite systems.

Here is the simplest example I can think of. Start with a massive body, call it a star, alone in an empty universe. Then add another massive body, a planet, so that it is captured by the star's gravity. Now we have a new concept, that of an orbit, and empirical regularities that can be expressed as Kepler's Laws. More is different in the sense that if you only have one body, you have neither orbits nor laws. And with its laws, the system satisfies Laughlin's definition. If the definitions apply to this simple system, then they apply to everything. More is always different. The definitions fail to separate a class of emergent phenomena from a class of non-emergent phenomena.

This also highlights the weakness of Israeli and Goldenfeld's definition. How many bodies constitute a "large assemblage"? If 2 bodies don't count, do 3 (in which the system becomes chaotic, as we'll see in the next chapter)? Is a solar system enough? A galaxy? Or do we need the trillions of components in the systems studied by condensed matter physics? In terms of the Game of Life, is a 10×10 grid enough, 100×100 or does it have to be larger? There's no way of answering these questions which isn't arbitrary so this definition also fails to distinguish between emergent and non-emergent systems.

Finally, theory and experiment are far more intertwined than Laughlin claims. Sometimes theory comes first. Examples are Maxwell's discovery of electromagnetic radiation (see the next chapter) or the prediction of the Higgs boson. Such predictions are not limited to high energy physics. A 2021 review notes that "…the literature seems replete with predictions of and about 2D materials".[6]

6.4   Physicist vs Physicist

Anderson writes that his article

---

[6] Penev et al. (2021).

...was unquestionably the result of a buildup of resentment and discontent on my part and among the condensed matter physicists I normally spoke with...[7]

It was squarely directed against the "arrogance of particle physicists" who think "if everything obeys the same fundamental laws, then the only scientists who are studying anything really fundamental are those who are working on those laws".[8] Laughlin seems to have had a similar experience:

> One common response in the early stages of learning is that superconductivity and the quantum Hall effect are not fundamental and therefore not worth taking seriously.[9]

Despite all this, neither article gives a convincing reason to reject the existence of fundamental things. Fundamentality means generality. Kepler's Laws only explain one set of phenomena. Newton's Law of Gravitation explains many. In the same way, Laughlin's model of the fractional quantum Hall effect uses quantum physics but you'd be hard put to use his model to explain the orbitals of a hydrogen atom. Here is Stephen Weinberg (Nobel Prize for Physics, 1979)

> ...by elementary particle physics being more fundamental I do not mean that it is more mathematically profound or that it is more needed for progress in other fields or anything else but only that it is closer to the point of convergence of all our arrows of explanation.[10]

Is such a preference for generality mere ideology? If so, it is an ideology at the heart of science.

The whole debate can often seem a squabble over who gets to call their work "fundamental". What's makes this more puzzling is that high energy physics (the term particle physics has fallen out of fashion) and condensed matter physics seem in many ways two sides of the same coin. Symmetry breaking is central to both, as are the techniques of effective theories and renormalization. In Chap. 2, I mentioned that virtual photons acquire mass in superconductors. The mechanism behind this is closely linked to the Higgs field, part of the standard model of high energy physics. Indeed, it is sometimes called the Anderson-Higgs mechanism.

---

[7] Anderson (2001).
[8] Anderson (1972).
[9] Laughlin (1999).
[10] Weinberg (1992), p. 55.

6.5   A Straw Man

The first paragraph of his article shows that Anderson is a card-carrying reductionist.

> The reductionist hypothesis may still be a topic for controversy among philosophers, but among the great majority of active scientists I think it is accepted without question. The workings of our minds and bodies, and of all the animate or inanimate matter of which we have any detailed knowledge, are assumed to be controlled by the same set of fundamental laws, which except under certain extreme conditions we feel we know pretty well.[11]

Laughlin is much less keen, I've already a cited a passage in which he lauded the end of the age of reduction. But his criticism of reductionism seems to be directed against a caricature. This portrays reductionism as being about tearing systems apart until their components are isolated, trying to explain the behaviour of the whole from these isolated components and asserting that the only interesting questions are at this lowest level. Such caricatures are surprisingly widespread in discussions of reductionism.[12]

To see how much of a straw man this is, you need to look no further than the two-body gravitational problem. It simply makes no sense to try to explain the motion of the planet without describing its interactions with the star. A reductive explanation involves explaining the behaviour of a whole in terms of its parts and this necessarily involves describing interactions among the parts. Here's Alex Rosenberg's definition of reduction in the context of biology:

> Reductionism is the thesis that biological theories and the explanations that employ them do need to be grounded in molecular biology and ultimately physical science, for it is only by doing so that they can be improved, corrected, strengthened, made more accurate and more adequate, and completed.[13]

When I return to the examples in Chap. 15, I will discuss how they can be explained reductively. For all of them, from quarks up to Churchill's nose, I think such explanations play the role that Rosenberg describes, improving and strengthening our understanding.

---

[11] Anderson (1972).

[12] An interesting discussing of the ways in which the term reductionism is used can be found in Riesch (2015).

[13] Rosenberg (2006), p. 4.

This straw man of reductionism will play a role in the discussion of contextual emergence in Chap. 8. In Chap. 13, I'll discuss how some opponents of reductionism not only direct their critique against a straw man but also treat it as if it were the bogeyman.

6.6   Discussion

More is different fails as a definition of emergence. Physics explains the behaviour of the parts and the interactions between the parts which constitutes the behaviour of the whole. Many-body systems are described by an expression called a Hamiltonian. This represents the total energy of the system as a sum over the kinetic and potential energy of its component particles. Interactions between particles means that in general the Hamiltonian will depend on the position and momentum of every particle. Also, in general, the energy of every particle will depend on the position and momentum of all the others. Take a many body system, add an extra particle and in general the energy of every other particle will change. This standard physics tells us why more is different and why more is always different.

As a way of clarifying this, let's take a hypothetical case in which "more is different" would have force as a definition. Start with a crystal lattice made up of atoms bonded together into a repeating geometric pattern. If you progressively remove atoms from the lattice, it will at some point start to fragment and will finally lose its lattice-like character all together. This process of disintegration will be gradual.

Now imagine that the force which holds the lattice together only switches on when you have more than N atoms. Below N, there's no force so no lattice. Above N, the force causes the lattice to form. There is a sharp threshold where behaviour changes radically. One Hamiltonian describes the system up to N then a different one takes over. Such a *configurational force* was proposed by the British Emergentists in the 1920s. More is different would then take on a specific meaning. There's no evidence that such forces exist, but if they did they would be a case of strong emergence involving something beyond current physics.

The passage I cited from Anderson ended with the words "…at each level of complexity entirely new properties appear, and the understanding of the new behaviors requires research…". I think this is the main sense in which the term emergence is used in condensed matter physics. There are interesting questions at every scale and there is no reason to give questions at the lowest level priority.

6.7   Further Reading

Humphreys (2014) is a rare discussion of Anderson's article from a philosophical perspective. Weinberg (1992), Chap. 3 gives a reductionist's view and Anderson's autobiography, Anderson (2011) his angle along with lots of fascinating background. Laughlin (1999) is a good introduction to his viewpoint as is his book, Laughlin (2005). Coleman (2019) is an interesting discussion of the relation between reduction and emergence in condensed matter physics. On the relation between the Higgs effect and superconductivity, see Anderson (2015). For configurational forces and a general discussion of British Emergentism, see McLaughlin (2008).

More suggestions for reading can be found at www.TheMaterialWorld.net.

References

Anderson PW (1972) More Is Different: Broken symmetry and the nature of the hierarchical structure of science. Science 177:393–396. https://doi.org/10.1126/science.177.4047.393

Anderson PW (2001) More is different - one more time. In: More is different: fifty years of condensed matter physics. Princeton university press, Princeton (N.J.)

Anderson PW (2011) More and different: notes from a thoughtful curmudgeon. World Scientific, New Jersey

Anderson PW (2015) Higgs, Anderson and all that. Nature Phys 11:93–93. https://doi.org/10.1038/nphys3247

Coleman P (2019) Emergence and reductionism: An awkward Baconian alliance. In: Gibb SC (ed) The Routledge handbook of emergence. Routledge, New York

Humphreys P (2014) More is Different…Sometimes: Ising Models, Emergence, and Undecidability. In: Why more is different. Springer, New York

Israeli N, Goldenfeld N (2006) Coarse-graining of cellular automata, emergence, and the predictability of complex systems. Phys Rev E 73:026203. https://doi.org/10.1103/PhysRevE.73.026203

Laughlin RB (1999) Nobel Lecture: Fractional quantization. Rev Mod Phys 71:863–874. https://doi.org/10.1103/RevModPhys.71.863

Laughlin RB (2005) A different universe: reinventing physics from the bottom down. Basic Books, New York

Laughlin RB (2021) Is Emergence Fundamental? https://www.youtube.com/watch?v=qT9iDcajqMo. Accessed 11 Jun 2024

McLaughlin BP (2008) The rise and fall of British emergentism. In: Bedau M, Humphreys P (eds) Emergence: contemporary readings in philosophy and science. MIT Press, Cambridge, Mass

Ong NP, Bhatt RN (2001) More is different: fifty years of condensed matter physics. Princeton University Press, Princeton (N.J.)

Penev ES, Marzari N, Yakobson BI (2021) Theoretical Prediction of Two-Dimensional Materials, Behavior, and Properties. ACS Nano 15:5959–5976. https://doi.org/10.1021/acsnano.0c10504

Riesch H (2015) Reductionism as an Identity Marker in Popular Science. In: Wagenknecht S, Nersessian NJ, Andersen H (eds) Empirical philosophy of science introducing qualitative methods into philosophy of science. Springer, Cham

Rosenberg A (2006) Darwinian reductionism, or, how to stop worrying and love molecular biology. University of Chicago Press, Chicago

Weinberg S (1992) Dreams Of A Final Theory: The Search for The Fundamental Laws of Nature. Vintage Books, New York.

7

Weak Emergence: When You Can't Do the Maths

Summary Some systems can be solved to give closed-form solutions. Some can only be simulated. Simulation emergence defines the latter as weakly emergent. However it is not a useful definition. If we take it literally it applies to everything, since only the most idealised systems have closed-form solutions. If we take it more loosely, it collapses into a tangle of borderline cases. I illustrate this with a discussion of the Game of Life. The importance the definition places on closed-form solutions raises the fascinating question of why maths is treated as the natural way of describing the world.

At the start of Chap. 5, I gave the example of a pendulum as a system that has a closed-form solution. A few lines of algebra give a simple equation which describes its motion. If you know the starting position, you can use this equation to calculate the position at any time in the past or the future. Contrast this with the Game of Life of Sect. 2.11. The only way to find out what is going to happen is to take the starting state and run the system forwards in time. There is no shortcut, no closed-form solution. This is simulation emergence. In this sense, the Game of Life is weakly emergent; the pendulum isn't.

The chapter starts by explaining simulation emergence and discussing which of the examples of Chap. 2 it can be applied to. However it turns out that it is not a useful definition. If we take it strictly, the definition applies to everything. If instead we take it more loosely, it gets lost in a tangle of borderline cases. A further problem is that it is contingent on the state of

© The Author(s), under exclusive license to Springer Nature
Switzerland AG 2025
L. Graham, *Physics Fixes All the Facts*, The Frontiers Collection,
https://doi.org/10.1007/978-3-031-69288-8_7

scientific knowledge. New techniques will remove phenomena from the category of simulation emergence. Finally, the definition can be seen as part of a much wider fetishisation of maths. This leads to the fascinating question of why maths describes the universe at all.

7.1   Simulation Emergence

We've already seen (Sect. 5.1) the distinction between solving a system to obtain a closed-form solution and studying it by simulation. In an influential body of work starting with a paper[1] in 1997, Mark Bedau defines a system as weakly emergent if it has no closed-form solution. His formal definition is

> Macrostate P of [microstate] S with microdynamic D is weakly emergent [if and only if] P can be derived from D and S's external conditions but only by simulation[2]

You can take macrostate to mean a high level state, and microdynamic as what is going on at a low level. Before looking at an example, some general points. Firstly, Bedau emphasises that this definition is compatible with reductionism: if you could know the low level state of a system exactly, you could use a simulation based on the laws of physics to predict its evolution. Secondly, this definition has nothing to do with causality. In the language of Chap. 3, it is epistemic not ontological. Thirdly, a point of terminology: the absence of a closed-form solution is also referred to as *non-deducibility* or *non-derivability*.

To see what this means in practice, let's again take the basic problem of celestial mechanics: the description of bodies orbiting each other under gravity. Newton's Law of Gravitation states that two bodies experience a gravitational force proportional to the product of their masses and inversely proportional to the square of the distance between them. For a system consisting of two bodies, it is easy to write down the force experienced by each of them. Then a few lines of algebra give the closed-form equations of motion for the two bodies.[3] Choose some initial positions and, just as for the pendulum, you can calculate the exact positions of the bodies forward into the infinite future or backwards into the infinite past. Then sit back and listen to this music of the spheres.

---

[1] Bedau (1997).

[2] Bedau (1997). A similar definition is in Darley (1994).

[3] See a classical mechanics textbook such as Barger and Olsson (1995).

For three bodies, the force experienced by one of them is the sum of the gravitational attraction of the other two. But despite this simplicity, the system can only be solved explicitly for restricted special cases. The most useful one is when one of the bodies has negligible mass compared to the other two, for example a spacecraft travelling through the earth-moon system or an asteroid influenced by the gravity of Jupiter and the sun. Others are when the orbits are circular; when the bodies are at the vertices of an equilateral triangle or when the three bodies are of equal masses moving in a figure of eight.

In general there is no closed-form solution. But the system is easy to solve by simulation. When this is done, the result is mesmerising patterns of motion with long periods of smoothness punctuated by abrupt changes. An infinitesimal change in the initial conditions can produce completely different patterns. Such simulations have also discovered thousands more stable configurations, some of which are illustrated in Fig. 7.1.[4]

The three-body gravitational problem can be generalised to any number of bodies, a many-body problem, and used to model the solar system. We tend to think of the planets as following stable orbits for all time. But just as with the three-body problem, simulations show that over hundreds of millions of years the positions of the planets become completely unpredictable and there is the possibility of dramatic events. A 2009 article[6] found that small changes

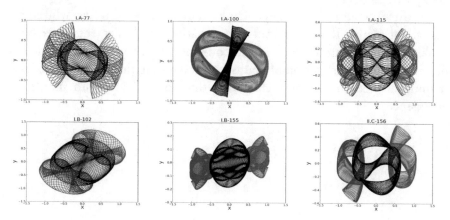

Fig. 7.1 Three body stable orbits[5]

---

[4] A link to animated versions of the figure is available at www.TheMaterialWorld.net.
[5] Source: Li and Liao (2017). https://link.springer.com/article/10.1007/s11433-017-9078-5. Reproduced with permission from Springer Nature.
[6] Laskar and Gastineau (2009).

in Mercury's orbit could lead to its colliding with the moon or Venus or even destabilising the whole system including the orbits of the giant planets.

Such behaviour is the hallmark of a chaotic system and simulation emergence is ubiquitous in such systems. Indeed, the definition is intended to distinguish chaotic behaviour from the calmer music of the spheres type of system. Bedau claims his definition is "metaphysically innocent, consistent with materialism, and scientifically useful".[7] Metaphysically innocent in the sense that all the causal powers reside at the lowest level and there is no downward causation. Consistent with physicalism since there is nothing but physics. Scientifically useful because, he argues, it gives formal structure to a concept used routinely by complexity scientists.[8]

The definition can be restated in terms of algorithmic incompressibility (Box 5.1). Returning to the example of celestial mechanics, the infinite trajectory of two orbiting bodies can be written as a few algebraic symbols: it is highly compressible. On the other hand, for the three-body problem the only way to study the evolution of the system is to let it run and see what happens. There is no short-cut: the system is algorithmically incompressible.

Which of the examples of Chap. 2 are simulation emergent? The answer is all of them. From quarks to the atom at the tip of Churchill's nose there are no exact closed-form solutions. Even for the hydrogen atom, where a closed-form solution is written in the top right corner of Fig. 2.3, an approximation is required to obtain it.

The first question that Bedau's definition bring to mind is this: why all the fuss about maths? If what matters is explanation, the "emergent" three body problem is as well understood as the non-emergent two-body problem. The question of this fetishisation of maths is so interesting that I'm going to leave it until last. Let's begin by highlighting three other issues with simulation emergence.

---

[7] Bedau (1997), p. 376.

[8] Bedau also argues that his definition explains how high level phenomena have explanatory autonomy and hence how explanation in terms of the higher level can be useful. This depends on a property called multiple realization which is the subject of the next chapter.

7.2   Which Systems Have Closed-Form Solutions?

The answer is: not many. The laws of nature are astonishingly simple (more on this in Sect. 7.5). But we don't observe the laws directly, we see their outcomes. And these outcomes turn out to be far more complex than the laws that underlie them.

Only the most idealised systems have closed-form solutions. Even the two-body gravitational system I described requires the assumption that the bodies are perfectly spherically symmetrical so can be treated as point masses. The equations of quantum mechanics can be solved exactly for simple systems, Wikipedia provides a list of around 30. However, for the two-body hydrogen atom, a closed-form solution is only possible by ignoring interactions between the electron and vacuum fluctuations (the spectral consequence of which is known as the Lamb shift). It's worth reflecting on this. Putting general relativity aside, for a physicalist everything is quantum physics. Yet there are no exact solutions for real systems. This means that simulation emergence is ubiquitous. And this means that it is not a useful definition since it fails to exclude anything.

Given the absence of closed-form solutions, you may ask why physics, and indeed science in general, is so full of what looks like closed-form solutions. The answer is that, like the solution to the hydrogen atom, these solutions involve some sort of approximation. This is the topic of the next section.

7.3   Borderline Cases

Scientists tend to do whatever it takes to understand the system they are interested, simplifying and approximating as necessary. There are three broad approaches. The first is to solve systems numerically. The second is to approximate the equations describing the system and then solve this approximate system, usually numerically though there may be closed-form solutions. The third is to use an approximate theory. Let's look at each of these in turn.

The closed-form solution for a pendulum is only valid for small oscillations. In general, the motion depends on an elliptic integral with no closed-form solution. However, the integral is well-understood and can be numerically computed to arbitrarily high accuracy. Solutions will be approximate, but the size of the approximation is no more than a matter of the computing power used.

Such numerical solution methods are ubiquitous. Let's say we want to find the roots of a cubic equation. Who can ever remember the formula to solve them? Instead, use a program to plot the curve. We can then see approximately where it crosses the axis and zoom in to plot the curve at a higher resolution in the vicinity of these crossings to improve the accuracy. This can be continued right to the limit of machine precision, or beyond with some tricks. There may be no closed-form solution, but we can know the system to arbitrary accuracy. Does this count as simulation emergence?

Now to approximating the equations that characterise the system. The non-linear systems of equations that describe most physical systems are hard to solve. But it is much easier to solve linear systems and it is easy to transform a non-linear equation into a linear one. You need to choose a point around which to linearise and the closer you are to this point the more accurate the solution will be. If you want to increase the accuracy, you can add in higher order terms. Does this count as simulation emergence?

Figure 7.2 shows an example. Say the system you're interested in is, like the pendulum, described by a sine function. Now imagine you don't know how to calculate this function but you do know how to expand it. This is of course highly artificial, but bear with me. On both panels of the figure, the solid line shows the sine function. I've chosen to expand around its peak, marked by the dashed vertical line.[9] Both panels have the same x-axis range. The peak is clearer on the right panel simply because the range of the y-axis is much smaller; you can just about see the peak on the left panel. The other lines show increasingly good approximations, ranging from first order (a straight line) to second order (a cubic) and so on. As you keep adding in higher order terms, you get closer and closer to the exact value. In Chap. 10, I will discuss a more realistic example in the context of the Renormalization Group Transformation.

Then there are approximate theories. For real systems, there are often a whole range of high level theories which predict behaviour to a high degree of accuracy. Examples are found in every area of physics: fluid dynamics (Navier–Stokes equations); ionic solutions (Debye–Hückel equation) and non-ideal gases (van der Waals equation). In Chap. 10, we'll see all physical theories involve approximations at some level. Does solving a system using these methods count as simulation emergence?

If you accept that these approximations count as closed-form solutions in the definition of simulation emergence, you run into another problem.

---

[9] The expansion is $\sin x = x - \frac{x^3}{3!} + \frac{x^5}{5!} - \frac{x^7}{7!} + \cdots$ So a linear approximation is $\sin x \approx x$, a second-order approximation $\sin x \approx x - \frac{x^3}{3!}$ and so on.

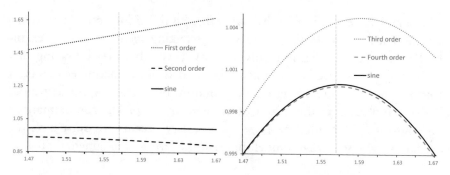

Fig. 7.2 Approximate solutions

With strong enough assumptions, you can model anything with a closed-form solution. Do you want to know the ultimate model of the universe, which carefully takes account of the interaction of every field and particle since the big bang to come up with a relation between the universe's age (t) and its size (r)? To a first-order approximation, it's $r = kt$ where k is a constant given by the current radius of the universe divided by its age. That's a closed-form solution. You can write down a similar equation for any other system. If everything can be approximated and approximations are not simulation emergent, then nothing is.

Simulation emergence is based on the distinction between two solution methods. As I discussed in Sect. 5.1, there is a third: statistical methods applied to many-body systems. Let's take the standard example of a litre of air in a closed well-insulated box. Although the underlying laws of motion are effectively Newtonian, the box contains around 10^{22} atoms so cannot be solved either exactly or by simulation. Statistical methods work by relying on the law of large numbers to obtain average properties of systems in equilibrium. And these are extraordinarily accurate. For example the temperature fluctuations at equilibrium for our box of gas are of the order of 10^{-10}K.[10] There are few other physical theories that produce such accurate results. Where do these statistical methods fit in the definition?

To sum up, one of the attractions of simulation emergence is that it appears to be scientifically useful in that it gives a clear way of distinguishing a set of phenomena as emergent. However these borderline cases show that there is no such clear distinction and, worse, the borderline cases seem to encompass most of what scientists do.

---

[10] The size of fluctuations is given by $\Delta T = T\sqrt{\frac{k}{C_v}}$ where Boltzmann constant k = 1.4×10^{-23} J/ K and the isochoric specific heat capacity of air C_v is 21 J/mol/K.

Bedau might respond to this by saying that his definition applies to systems that are best simulated using models of interacting agents[11] and the examples I've given do not fall into this category. I would answer by saying that all physical systems are best simulated by models of interacting elements. A proton is made of interacting quarks; an atom of interacting nucleons and electrons; molecules of interacting atoms and so on up. The approximations that constitute the borderline cases are necessary because we don't have the computational resources to model things at the level of fundamental physics.

7.4   The History of Closed-Form Solutions

Whether a closed-form solution exists or not depends on the techniques available. This doesn't mean it is entirely subjective since we can take it to be given by the best-available knowledge. But it does means that the scope of simulation emergence will change over time.

Before Galileo and Newton, there were no such solutions. Their expansion is linked to the progress of science. In 1926, Erwin Schrödinger published his revolutionary solution for the hydrogen atom.[12] The following year, it was applied to the hydrogen molecule and in 1955 to the nitrogen molecule.[13] Almost a century later, quantum physics is applied, albeit with approximations, to a vast range of phenomena. In response to this, Bedau asserts that his definition is ontological and not epistemic:

> If a genius like Newton discovers a new short-cut derivation for macro properties in a certain class of system, this changes what properties we think are weakly emergent but not which properties are weakly emergent.[14]

Elsewhere, he claims that algorithmic incompressibility is also ontological.[15] However it's all very well saying that some things are ontologically simulation emergent. But, in the absence of a formal mathematical proof, something which is vanishingly rare for systems of any complexity, we can never know what these things are.

The definition means that over the history of science the scope of what we know to be simulation emergent narrows and can be expected to narrow

---

[11] Bedau (2008), p. 447.
[12] Schrödinger (1926), for a translation see Schrödinger (1982).
[13] Heitler and London (1927), Scherr (1955).
[14] Bedau (2002), p. 16.
[15] Bedau (2008).

further in the future as our understanding and techniques improve. Simulation emergence does not tell us anything useful about the world but is about the models that scientists use. More on this in Chap. 10.

7.5   The Language of Nature?

Why should we make a fuss about whether there is a closed-form solution or not? The answer goes right back to the start of modern science. Here's Galileo:

> Philosophy is written in this grand book—I mean the universe—which stands continually open to our gaze, but it cannot be understood unless one first learns to comprehend the language and interpret the characters in which it is written. It is written in the language of mathematics, and its characters are triangles, circles, and other geometrical figures, without which it is humanly impossible to understand a single word of it; without these, one is wandering about in a dark labyrinth.[16]

The mention of geometrical figures situates Galileo in the long history of sheer wonder in the face of the ability of maths to explain the world. Maths has been a key part of the scientific project ever since and closed-form solutions capture the idea of simplicity and beauty that motivates many scientists. This banal statement contains something extraordinary: why should the physical world be describable by maths?

In 1960, Eugene Wigner, who three years later won the Nobel Prize for Physics, wrote an influential paper with the title "The Unreasonable Effectiveness of Mathematics in the Natural Sciences".[17] The article starts with an anecdote about the constant π. We first encounter it is the ratio of the circumference of a circle to its diameter. But as we learn more maths it pops up everywhere: in the Gaussian distribution (which, thanks to the Central Limit Theorem, describes large numbers of pretty much anything), in Heisenberg's uncertainty principle, the Schrödinger equation and Einstein's field equations (which between them, if we are good physicalists, describe everything). Then there is Euler's magnificent identity[18] which relates π to four other mathematical constants.

---

[16] Galileo (1960), p. 183.
[17] Wigner (1960).
[18] $e^{i\pi} + 1 = 0$.

Wigner says his question of the unreasonable effectiveness of maths has two parts. First, why there are simple laws of nature. Second, why maths describes them. As an example of the first, he gives the example of a regularity that Galileo discovered: if you drop two objects of different weights they hit the ground of the same time. What is extraordinary is that this regularity holds everywhere and is independent of the myriad of things that could logically affect it (the colour of the rock, whether it's night or day, whether the person dropping the rock is left or right-handed). Without such independence, the sort of controlled experiments on which science relies would be impossible. The question of why the laws of nature are simple is fascinating but not one that it relevant here.

Granted that there are simple laws, why can they be described by maths? Maths is usually created by mathematicians doing what they do, playing with abstract formal systems to satisfy their sense of mathematical beauty without a thought about the physical world. Then, decades later, it turns out that one of these abstract systems is exactly what a physicist is looking for to describe their pet theory. The most famous example is the theory of Hilbert spaces, developed during the first decade of the 20th then, from the late 1920s, providing one of the most elegant formulations of quantum theory.

What's more, there are often several different mathematical formulations of the same physics. Richard Feynman describes three ways in which Newtonian gravitation can be formulated.[19] A 2002 paper[20] lists twelve distinct formulations of classical mechanics and nine of quantum mechanics. And maths does not just describe reality but can lead to discoveries. In the mid-nineteenth century Maxwell added a term to one of the equations of electrodynamics to make it consistent with the conservation of electric charge. The extra term, called displacement current, was added by mathematical analogy with the Newtonian principle of conservation of mass, and had nothing to do with empirical work. Maxwell then showed the modified equation predicted the existence of electromagnetic radiation propagating at the speed of light.

The only systems for which it is evident that maths will be useful are those containing large number of similar parts. This is the realm of statistical mechanics. In such cases, averaging will give the aggregate properties of the system. You might expect that the results of such an averaging process

---

[19] Feynman (1990).
[20] Styer et al. (2002).

would be devoid of physical content. But among them is a law of nature, the second law of thermodynamics.[21]

What can we make of all this? There are any number of potential explanations. At one end of the spectrum is the Max Tegmark's Mathematical Universe Hypothesis, which states that the universe is itself a mathematical structure.[22] At the other is the idea that maths is species specific but that natural selection tunes our brains to be suited to mathematical tools that are useful to describe the world we evolved in.[23]

The answer I find most satisfying is due to the physicist Dennis Dieks. Maths investigates the consequences of simple axioms. Dieks argues that the laws of physics are just such simple axioms so maths is the same sort of project as physics:

> ...the very observation that mathematics has no physical content can take away most of the surprise. Indeed, exactly because mathematics is a 'freely floating construction', not tightly bound to sense experience, it is extremely flexible and versatile—and therefore useful.[24]

What mathematics as a discipline does is constantly churn out new formalisms and it should come as no surprise that some of these are applicable to physics. This is supported by two observations. On the one hand, there are many areas of maths with no application to physics. On the other, physicists sometimes have to invent maths to explain the systems they are interested in. The most famous example of this is Heisenberg devising a new algebra for his seminal 1925 paper on quantum theory. When Max Born read it he realised that Heisenberg had reinvented matrix algebra which was at the time mostly unknown to physicists.[25] A less well known case is that of Pascual Jordan who in the early 1930s similarly invented a new algebra to describe quantum theory. Jordan algebras didn't catch on with physicists but have been used widely by mathematicians.[26]

To return to simulation emergence, it strikes me that it is grounded in the sense that using the language of nature directly is somehow more

---

[21] It turns out that implicit in the averaging is the assumption that fundamental physics displays time-reversal symmetry. Without this, systems would not tend to a state of maximum entropy and the universality of the second law would be lost. For a fascinating discussion, see Strasberg (2024).

[22] Tegmark (2008).

[23] Steiner (1998), p. 50, see also Weinberg (1992), p. 132.

[24] Dieks (2005), p. 116.

[25] Bernstein (2005).

[26] Dahn (2023).

fundamental than doing simulation. Yet is it? Unless Tegmark's Mathematical Universe Hypothesis is true, the universe manages to do everything it does without any maths. With this perspective, physical laws are low-dimensional approximations to the mysterious way in which the universe works. If what we care about is explanation and understanding, there is no reason to take a closed-form explanation as somehow better than one arising from a simulation.

7.6   Example: The Game of Life

Cellular automata in general and the Game of Life in particular, are canonical examples of the idea of simulation emergence. Their simple rules give rise to intricate dynamic patterns and it is rare to find an author who can resist a breathless description of his favourite.

What's particularly interesting about the Game of Life is that there is a proof that it is algorithmically incompressible. The first step is to show that patterns exist which function as logic gates and that they can be assembled in arbitrary ways. This, along with a few other tricks is sufficient to show that the Game of Life is Turing-complete i.e. it can do anything a Turing machine can. This is not just a theoretical result. In 2000, Paul Rendell created a Turing machine in the Game of Life i.e. a pattern that can carry out the functions of a Turing machine.

In 2010, he extended this to create a universal Turing machine. Then the halting problem applies. In the context of the Game of Life, this means that for a given starting pattern there is no way in general proving whether a given pattern will ever be created. The only way to find out is to run the system and see. In this strong sense, the Game of Life is algorithmically incompressible.

But this result is not as important as it may seem. Proving some things are non-computable does not stop many things being computable (Sect. 5.4). In the same way, proving that the Game of Life is globally incompressible does not exclude that there may be limited compressibility and explanations of specific patterns.

What does this mean in practice? It turns out there are empirical laws in the Game of Life. One of the simplest is that "gliders move in straight lines until they collide with another live cell". You don't need to run the game to predict the motion of a glider, you just need to draw a straight line: this is compressibility. There are many other such regularities. Descriptions of patterns in the game are usually conjoined with an account of how the pattern evolves.[27]

A fascinating paper[28] extends this parallel. The authors note that when we simulate, or indeed observe, physical systems we do not have the ability to deal with every microphysical detail so instead take a coarse graining approach. I'll return to the idea of coarse graining in much more detail in the next chapter, but in the Game of Life it means some rule to aggregate adjacent cells into a single cell. A simple way of doing this would be to take a 3×3 block of cells and represent them by a larger cell which is alive is a majority of the 9 original cells were alive and dead otherwise. The paper shows that, if you are only interested in coarse grained information

> …undecidability and computational irreducibility are not good measures for physical complexity. Physical complexity, as opposed to computational complexity, should address the interesting, physically relevant, coarse-grained degrees of freedom. These coarse-grained degrees of freedom maybe simple and predictable even when the microscopic behavior is very complex.[29]

Simple are predictable are the antithesis of what Bedau seeks to capture with his definition of simulation emergence.

It seems to me that all this should make a hard-nosed physicalist jump for joy. The Game of Life shows how a mind-boggling range of complex behaviour can emerge from simple rules. I've mentioned the possibility of constructing universal Turing machines. If this wasn't enough, there are also Universal Constructors[30] and Sect. 2.11 mentioned the pattern which makes copies of itself. Does this process have any limits? Conway's classic article concludes:

> It's probable, given a large-enough Life space, initially in a random state, that after a long time, intelligent self-reproducing animals will emerge and populate some parts of the space.[31]

The important point is that because the underlying rules are simple, however complex the behaviour, there is always an exact, straightforward

---

[27] Bedau might argue that such regularities are emergent in the sense that they can only be discovered by running the Game. But this applies to everything. Laws of nature are empirical regularities, not the result of a priori deductions.

[28] Israeli and Goldenfeld (2006).

[29] Israeli and Goldenfeld (2006).

[30] https://conwaylife.com/wiki/Universal_constructor.

[31] Conway (2004).

reductionist explanation of every step. So straightforward, that you can do the necessary calculations with a sheet of squared paper and a pencil (as long as your paper and lifespan are large enough!). If such simple rules can lead to such complex behaviour, it seems much les surprising that the more complicated rules of the physical universe can lead to life, intelligence and consciousness. And that, just as in the Game of Life, there is always a crystal-clear reductionist explanation if you know how to find it.

7.7   Related Definitions

There are two other definitions which are closely related to simulation emergence. The first, which I already mentioned in Sect. 3.12 categorises weakly emergent phenomena as unexpected. Take the Game of Life. Given the rules, no-one would have expected the existence of gliders let alone Turing machines. But once they have been discovered, it is easy to show that are direct consequences of the rules.

The second is by Eleanor Taylor:

> Given components A, B, C… n arranged in relation r into a whole, and an observer O, property x of the whole is emergent for O [if and only if] there is **no scientific explanation** available to O of the fact that the following regularity obtains of natural necessity: Whenever components A, B, C…n are combined in relation r, the resulting whole instantiates property x.[32]

Taylor goes on to say that she intends to include "…failures of deducibility and derivability, as explanatory failures." So her definition nests simulation emergence. This strikes me as peculiar since, as with the 3-body problem or the Game of Life, there are impeccable reductionist explanations of everything.

Both these definitions share the feature that the more physics we know, the fewer phenomena seem emergent. So the boundaries of what is emergent change over time as science progresses.

---

[32] Taylor (2015), my emphasis.

7.8 Discussion

Closed-form solutions are limited to the most idealised systems. If you want to call other systems simulation emergent, there's nothing stopping you. But then everything in the world is emergent. On the other hand, simply discard the term and you lose nothing.

Bedau intended his definition to separate simple from complex behaviour. But it fails even to do that. There are systems which show aspects of chaotic behaviour but have closed-form solutions.[33] And there are systems, such as the large-angle pendulum discussed in Sect. 7.3, which don't have closed-form solutions but don't show chaotic behaviour.

In Chap. 5, I introduced the Church-Turing-Deutsch principle which states that we can simulate all physical systems. Bedau tells us that "...the [simulation] emergence perspective is ontologically and causally reductionistic".[34] Combining these means that simulations can give us reductionist explanations of all physical systems. We may need simplified models for our imaginative understanding and maths often plays an important role in the construction of such models. But for the representational understanding, which is the goal of science, simulation may be enough.

7.9 Further Reading

For a discussion of simulation emergence from a philosophical perspective, see Wilson (2021), Sect. 5.2.1. For a history of solutions to the Schrödinger equation, see Esposito and Naddeo (2013). For a general overview of the role of maths in science, see Dorato (2005) and for a criticism of the Mathematical Universe Hypothesis see Piccinini and Anderson (2018). Lloyd (2007), Chap. 8 shows how simple rules can lead to complex outcomes. A thorough discussion of Turing machines in the Game of Life is in Rendell (2016) and more details of the computational irreducibility of the Game of Life in Zwirn and Delahaye (2013).

More suggestions for reading can be found at www.TheMaterialWorld.net.

---

[33] Faghani et al. (2019).
[34] Bedau (2002), p. 43.

References

Barger V, Olsson MG (1995) Classical mechanics: a modern perspective. McGraw-Hill, New York

Bedau MA (1997) Weak Emergence. Philosophical Perspectives 11:375–399. https://doi.org/10.1111/0029-4624.31.s11.17

Bedau MA (2002) Downward Causation and the Autonomy of Weak Emergence. Principia: An International Journal of Epistemology 06:5–50

Bedau MA (2008) Is Weak Emergence Just in the Mind? Minds & Machines 18:443–459. https://doi.org/10.1007/s11023-008-9122-6

Bernstein J (2005) Max Born and the quantum theory. American Journal of Physics 73:999–1008. https://doi.org/10.1119/1.2060717

Conway JH (2004) What is Life? In: Winning ways for your mathematical plays. Peters, Wellesley, Mass

Dahn R (2023) Nazis, émigrés, and abstract mathematics. Physics Today 76:44–50. https://doi.org/10.1063/PT.3.5158

Darley V (1994) Emergent Phenomena and Complexity. In: Brooks RA, Maes P (eds) Artificial Life IV. The MIT Press, pp 407–412

Dieks D (2005) The Flexibility of Mathematics. In: Boniolo G, Budinich P, Trobok M (eds) The role of mathematics in physical sciences: interdisciplinary and philosophical aspects. Springer, Dordrecht, the Netherlands

Dorato M (2005) The Laws of Nature and the Effectiveness of Mathematics. In: Boniolo G, Budinich P, Trobok M (eds) The role of mathematics in physical sciences: interdisciplinary and philosophical aspects. Springer, Dordrecht, the Netherlands

Esposito S, Naddeo A (2013) The genesis of the quantum theory of the chemical bond. https://doi.org/10.48550/ARXIV.1309.4647

Faghani Z, Nazarimehr F, Jafari S, Sprott JC (2019) Simple Chaotic Systems with Specific Analytical Solutions. Int J Bifurcation Chaos 29:1950116. https://doi.org/10.1142/S0218127419501165

Feynman RP (1990) The character of physical law. MIT Press, Cambridge, Mass.

Galileo G (1960) The Assayer. In: The Controversy on the Comets of 1618: Galileo Galilei, Horatio Grassi, Mario Guiducci, Johann Kepler. University of Pennsylvania Press

Heitler W, London F (1927) Wechselwirkung neutraler Atome und homoeopolare Bindung nach der Quantenmechanik. Z Physik 44:455–472. https://doi.org/10.1007/BF01397394

Israeli N, Goldenfeld N (2006) Coarse-graining of cellular automata, emergence, and the predictability of complex systems. Phys Rev E 73:026203. https://doi.org/10.1103/PhysRevE.73.026203

Laskar J, Gastineau M (2009) Existence of collisional trajectories of Mercury, Mars and Venus with the Earth. Nature 459:817–819. https://doi.org/10.1038/nature08096

Li X, Liao S (2017) More than six hundred new families of Newtonian periodic planar collisionless three-body orbits. Sci China Phys Mech Astron 60:129511. https://doi.org/10.1007/s11433-017-9078-5

Lloyd S (2007) Programming the universe: a quantum computer scientist takes on the cosmos. Vintage Books, New York , NY

Piccinini G, Anderson NG (2018) Ontic Pancomputationalism. In: Cuffaro ME, Fletcher SC (eds) Physical perspectives on computation, computational perspectives on physics. Cambridge University Press, Cambridge New York, NY

Rendell P (2016) Turing Machine Universality of the Game of Life. Springer International Publishing: Imprint: Springer, Cham

Scherr CW (1955) An SCF LCAO MO Study of N2. The Journal of Chemical Physics 23:569–578. https://doi.org/10.1063/1.1742031

Schrödinger E (1926) Quantisierung als Eigenwertproblem. Annalen der Physik 384:361–376. https://doi.org/10.1002/andp.19263840404

Schrödinger E (1982) Collected papers on wave mechanics. Chelsea, New York, NY

Steiner M (1998) The applicability of mathematics as a philosophical problem. Harvard University Press, Cambridge, Mass

Strasberg P (2024) Why God plays dice: A pedagogical and accurate explanation of the second law. https://boltzmannsbrain.blog/why-god-plays-dice-a-pedagogical-and-accurate-explanation-of-the-second-law/

Styer DF, Balkin MS, Becker KM, et al (2002) Nine formulations of quantum mechanics. American Journal of Physics 70:288–297. https://doi.org/10.1119/1.1445404

Taylor E (2015) An explication of emergence. Philos Stud 172:653–669. https://doi.org/10.1007/s11098-014-0324-x

Tegmark M (2008) The Mathematical Universe. Found Phys 38:101–150. https://doi.org/10.1007/s10701-007-9186-9

Weinberg S (1992) Dreams Of A Final Theory: The Search for The Fundamental Laws of Nature. Vintage Books, New York.

Wigner EP (1960) The unreasonable effectiveness of mathematics in the natural sciences. Comm Pure Appl Math 13:1–14. https://doi.org/10.1002/cpa.3160130102

Wilson JM (2021) Metaphysical Emergence. Oxford University Press, Oxford

Zwirn H, Delahaye J-P (2013) Unpredictability and Computational Irreducibility. In: Zenil H, Franke HW, Wolfram S (eds) Irreducibility and computational equivalence: 10 years after Wolfram's A new kind of science. Springer, Heidelberg; New York

8

Weak Emergence: One from Many

Summary High level properties such as temperature, hardness or roundness are multiply realized: they are shared by systems with different microphysical makeups. Such properties are multiple realization (MR) emergent in the sense that they provide useful explanations independent of the details of their makeup. But MR emergence is everywhere and fails as a useful definition. It is everywhere because it is the nature of conceptual thought to abstract from detail. Studying multiply realized phenomena tells us much more about the nature of our cognition than it does about the nature of the world. Advocates of MR emergence assert that it prevents reduction. However the opposite is true, reductive physics explains how high level properties can be multiply realized.

Picture a ball rolling down a slope. Did you also picture what your ball was made of? There was no need to. Given some assumptions about gravity, friction etc., all balls behave in the same way. Any particular ball is a precise arrangement of a particular set of atoms, a specific microphysical makeup. Yet all of them share the property of rolling down a slope. This is *multiple realization*, high level phenomena can be realized by many different systems.

The coke machine of Sect. 3.11 was a more playful example. There is no question that each particular ball or coke machine is subject to physics. But the properties of rolling down a slope or dispensing a coke somehow float free of physics, they are multiple realization (MR) emergent.

Multiple realization is closely related to the idea of coarse graining and the chapter starts by clarifying these terms then looking at some examples

© The Author(s), under exclusive license to Springer Nature
Switzerland AG 2025
L. Graham, *Physics Fixes All the Facts*, The Frontiers Collection,
https://doi.org/10.1007/978-3-031-69288-8_8

taken from statistical mechanics. One of the most extraordinary phenomena is that phase transitions in different substances and systems share common properties. This feature, called universality, is a canonical example of MR emergence.

Next comes the critique. Once more, the definition fails because it applies to all high level properties. This is the case because multiple realization is the nature of conceptual thought. Our cognitive and perceptual limitations mean we can only think about systems by treating them as identical and making and often overlooking whatever approximations are necessary to do so. Multiple realization is an illusion since every physical system is distinct. Rather than multiple realization being a problem for reduction, the chapter ends by showing that the opposite is the case: reductive physics explains why different systems share similar high level properties.

8.1   Multiple Realization

Take your favourite special science and pick a concept from it. From astronomy, you might pick a solar system. From biology, a cell. From geology, an earthquake. From material science, brittleness. Whatever you chose, a vast number of different systems are covered by the concept. And each of these systems is a distinct arrangement of atoms, it is distinct at the microphysical level. This is multiple realization. The higher level concept is to some extent autonomous, robust to changes in some of its lower level details. This is MR emergence, a version of non-reductive physicalism that seems the mainstream view among philosophers.

Figure 8.1 shows a more abstract example. On the left side of the figure are 6 × 6 grids of cells which can be either on or off.[1] Then define a transformation which zooms out, mapping each of the 3 × 3 blocks of cells on the original grid onto a single cell in the 2 × 2 grids on the right side. If a majority of the 9 cells in the original block is on, the new cell is on, otherwise, it is off.

Start with the 6 × 6 grid at the top. The figure shows the mapping for two of its 3 × 3 blocks. In the one at the top left, four cells are on a five off so the new cell is off. In the one at the bottom right five are on and four off so the new cell is on. Then look at the 6 × 6 grid at the bottom of the figure. This has a different configuration from the first grid. But when we transform it using the same rule it gives an identical 2 × 2 grid.

---

[1] Such grids can be used to represent the Ising model of magnetism of which we'll see more in Chap. 10.

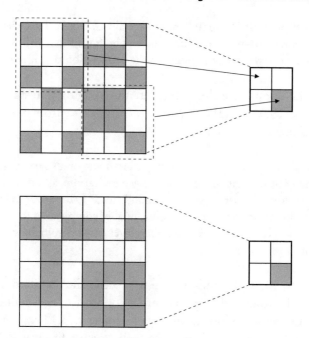

Fig. 8.1 Multiple realization

Now, if we start from the grids on the right side, they are identical. Since they correspond to different 6 × 6 grids, we can say they are multiply realized. The transformation I've described, this process of zooming out, is often called *coarse graining* since it moves from the fine grain of the microphysical to a coarser description.

Such coarse grained descriptions are MR emergent. Taken at face value, MR emergence seems to pose a problem for reduction. If a high level property corresponds to many low level arrangements, which one do you reduce it to? The link between low level and high level explanations has been broken. The attraction of this idea is obvious. We can be (apparently) hard-core physicalists and assert there is nothing to high level things but low level components, while at the same time asserting that reduction is not possible. We can be non-reductive physicalists.

The idea of multiple realization was first introduced by Hilary Putnam in the 1960s. He was interested in the question of whether mental states, such as pain, are identical with physical brain states. Putnam argues that this is at best extremely unlikely because that physical state must be:

> ...a possible state of a mammalian brain, a reptilian brain, a mollusc's brain (octopuses are mollusca, and certainly feel pain), etc. .... Even if such a state

can be found, it must be nomologically certain that it will also be a state of the brain of any extra-terrestrial life that may be found that will be capable of feeling pain before we can even entertain the supposition that it may be pain".[2]

Although he doesn't use the term, Putnam is saying that pain is multiply realized. In the 1970s, Fodor extended the argument from mental states to the special sciences in general.

Does anybody really doubt that mountains are made of all sorts of stuff? Does anybody really think that, since they are, generalizations about mountains-as-such won't continue to serve geology in good stead? Damn near everything we know about the world suggests that unimaginably complicated to-ings and fro-ings of bits and pieces at the extreme microlevel manage somehow to converge on stable macro-level properties.[3]

Subsequently, multiple realization has become the standard argument to use against reduction and the standard support for non-reductive physicalism. The definition of weak emergence in Sect. 3.11 was when only a subset of causal powers is relevant. MR emergence fits this definition. To see this, note that both you and I have read the previous sentence. Although our microphysical configurations are different, only some tiny subset of the microphysical possibilities in each of us is responsible for our both reading. Most of the differences between us are irrelevant.

8.2   Example: Statistical Mechanics

Statistical mechanics is a way of describing systems that are made from too many parts to keep track of. Instead of dealing with the microscopic details of the system, we can make do with aggregate variables. In this sense, it is the science of MR emergence. There are two key concepts. The microstate which is the microphysical configuration of the system. The macrostate which is the high level property we are interested in. Statistical mechanics explains the relation between them.

For a box full of air, the macrostate is what we choose to measure, temperature or pressure, for example, and the microstate relates to the details of its component molecules. For a room full of objects, the macrostate might

---

[2] Putnam (1967).

[3] Fodor (1997).

be some measure of tidiness, the microstate the position of the individual objects.

The central concept of statistical mechanics is entropy, which counts the number of microstates corresponding to a given macrostate. Conventionally, entropy is defined as the natural logarithm of this count. Returning to the example of Fig. 8.1, the macrostate is the 2 × 2 grid on the right side, the microstate is the 6 × 6 grid on the left. How many possible patterns of the 6 × 6 grid give the same 2 × 2 grid? The answer is around four billion.[4] This is the number of microstates that correspond to our chosen macrostate. So the entropy of each macrostate is around 22. If instead we mapped our 6 × 6 grid onto a macrostate of a 3 × 3 grid, each 3 × 3 pattern would have an entropy of 18. Take the extreme case where we map the 6 × 6 grid onto just one point. This would have an entropy of 24.

Thus entropy can be seen as a way of measuring multiple realization, at least for systems for which the micro and macrostates can be formally defined. The higher the entropy, the greater the degree of multiple realization.

8.3 Degrees of Freedom

There is something odd about MR emergence. It classifies things as emergent according to whether there are other, similar phenomena. But shouldn't emergence be an inherent and not a relative property? In 2010, Wilson proposed a closely-related definition of emergence which avoids this criticism,[5] called degrees-of-freedom (DOF) emergence.

The number of degrees of freedom is the smallest number of independent parameters necessary to fully describe a system. Take an example of a 2 × 2 grid of which the four elements can either on or off. This system has four degrees of freedom since to describe it we need four numbers. In general, an n x n grid will have n^2 degrees of freedom.

Wilson defines DOF emergence as follows. Take some high level system made up of lower level components. The high level system is emergent if we can describe its behaviour either with less degrees of freedom than the

---

[4] A 6 × 6 grid has 2^{36} configurations. A 2 × 2 grid 2^4. So each configuration of the 2 × 2 grid corresponds to 2^{32} configurations of the 6 × 6 grid. A 3 × 3 grid has 2^9 so each configuration corresponds to 2^{27} configurations of the 6 × 6 grid. A single point has just 2 configurations each of which corresponds to 2^{35} configurations of the large grid. To calculate entropy from these counts, take the natural logarithm.

[5] Wilson (2010).

low level system or with the same number of degrees of freedom but over a restricted range.[6]

Take another glance at Fig. 8.1. The 6×6 grids on the left side each have 36 degrees of freedom. When zoomed out they are mapped to 2×2 grids with four degrees of freedom. So the patterns on the right are DOF emergent. In contrast to MR emergence, we only need a single grid to see this.

All the systems studied by statistical mechanics are DOF emergent. Take a litre of a gas in a box. There will be around 10^{22} molecules each of which has six degrees of freedom (three for position and three for velocity). However the system can be described with just three variables, pressure, volume and temperature, one of which is redundant if the gas is ideal.

As a final example, let's return to the Game of Life[7] To specify one cell on the x–y grid requires three variables, two for the coordinates and one to specify whether it is on or off. Recall that a glider is a pattern of five cells (see Fig. 2.12). To describe it at one point in time requires 15 degrees of freedom. However, if we start the Game with just one glider and nothing else (to avoid collisions), we can describe the system at any point in time with 18 degrees of freedom (the 15 degrees of freedom that define the shape and position of the glider, two to describe the diagonal trajectory and one for time). If we want to describe the Game over 20 periods, instead of needing 15 degrees of freedom for each period, so 300 in total, we need just 18.

This fits the definition of weak emergence in Sect. 3.11. For subset of causal powers read subset of degrees of freedom. We have DOF emergence. This definition gives a strong sense in which reduction fails and, since it is applicable to single instances, avoids the objection I raised at the start of this section. It nests MR emergence, since if something is multiple realized it necessarily has reduced degrees of freedom. It also neatly includes contextual emergence which is the topic of the next chapter.

8.4   Universality

Phase transitions were one of the examples in Chap. 2. Different phases are usually separated by a phase boundary. However at some combination of pressure and temperature the phase boundary vanishes. This is known as the critical point. At higher temperature, the gas and liquid phases merge and the substance becomes a supercritical fluid.

---

[6] The full definition is in Appendix A.1.
[7] Adapted from Wilson (2021), p. 188.

Figure 8.2 is taken from a paper that measures the behaviour of various substances near their critical point. The x-axis shows the ratio of the density of the substance to that at the critical point. A value of unity means the liquid and the gas have the same density. Lower values represent the gas phase, higher values the liquid phase. On the y-axis is the ratio of the temperature to that at the critical point. A value of 1 means the substance is at the critical temperature. Values of critical temperature and density vary between substances; drawing the graph in terms of ratios allows behaviour around the critical point to be compared.

The points on the graph plot shows combinations of density and temperature at which liquid and gas coexist. This is done for eight different chemicals (neon, argon, krypton, xenon, nitrogen, oxygen, carbon monoxide and methane). What is quite extraordinary is that different substances with different molecular structures have effectively identical behaviour.

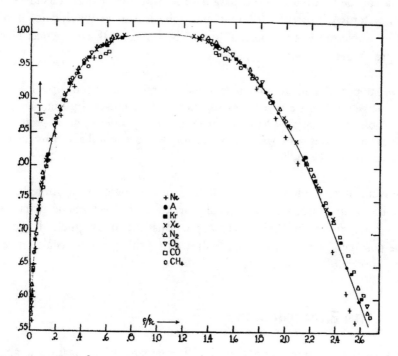

Fig. 8.2 Universality[8]

---

[8] Reproduced from Guggenheim (1945), https://doi.org/10.1063/1.1724033 with the permission of AIP Publishing.

The shape of the curve can be described by a single number, called a critical exponent[9] and the best fit line on the figure is drawn using a value of 1/3. Since all the points lie on this line, we can say that the critical exponent is the same across the eight chemicals. It turns out that this critical exponent is shared by magnetic materials. Experimental work from 1995 confirms that this holds for a ferromagnet over 18 orders of magnitude of temperature. What's more, this critical exponent is the same as drops out of a theoretical analysis of the Ising model (of which we'll see more in Chap. 10). It is extraordinary that the eight substances show the same behaviour. That an entirely unrelated system does so too is staggering.

This is the phenomenon of universality. Systems which share the same critical exponents, such as the 8 substances, ferromagnets and the Ising model, are said to belong to the same universality class. Wikipedia gives a list of 12 such classes. In each of these classes, the complexity of microphysical interactions can be reduced to a few parameters. And these parameters are shared across systems with vastly different microphysical makeups. It's not hard to see why universality is the poster-child of MR and DOF emergence. Here is Margaret Morrison:

> If we suppose that micro properties could determine macro properties in cases of emergence then we have no explanation of how universal phenomena are even possible. Because the latter originate from vastly different micro properties there is no obvious ontological or explanatory link between the micro-structure and macro behaviour.[10]

Universality in phase transitions is one example of a more general phenomena in condensed matter physics of high level phenomena being largely independent of their microscopic constituents. A quick glance back at the passages I quoted in Chap. 6 shows that this property is also implicit in the idea that "more is different".

8.5   The Examples

Which of the examples are MR or DOF emergent? Once again, the answer is all of them. They all have high level properties which, to some extent,

---

[9] This critical exponent is conventionally known as β and is given by: $\rho - \rho_c = |T - T_c|^\beta$, where ρ is density, T temperature and subscript c means the values at the critical point. There are various other critical exponents, see Stanley (1987), Chap. 3.

[10] Morrison (2014).

are independent of the precise details of their makeup. A proton's properties do not depend on exactly the combination of virtual particles that it contains.[11] The form of Churchill's nose is robust to changes in the configuration of its atoms. They all also involve limiting the degrees of freedom of their components. Universality is a more restrictive property, applying only to phase transitions and quasiparticles.

It's not just that these two definitions of emergence apply to all the examples, they apply to everything. Let's now look at why this is the case.

8.6   Concepts

The Oxford English Dictionary defines a concept as "…an idea of a class of objects, a general notion or idea". Here are the first lines of a book on the subject:

> Without concepts, mental life would be chaotic. If we perceived each entity as unique, we would be overwhelmed by the sheer diversity of what we experience and unable to remember more than a minute fraction of what we encounter. And if each individual entity needed a distinct name, our language would be staggeringly complex and communication virtually impossible. Fortunately, though, we do not perceive, remember, and talk about each object and event as unique, but rather as an instance of a class or concept that we already know something about.[12]

This is a description of multiple realization. Human thought is conceptual thought. If two things share a property, the property is multiply realized. If two things share a noun, that noun is multiply realized. Multiple realization is a property of our way of thinking, not of the world. Susan Langer wrote: "Our world 'divides into facts' because we so divide it".[13] That is why multiple realization applies to everything so is useless as a definition of emergence. It tells us nothing about the nature of the world, only about the nature of our cognitive processes.[14]

---

[11] Quantum indistinguishability is not relevant since multiple realization applies to properties, not the particles themselves. For example, both protons and neutrons have spin ½, despite their different compositions, so this property is multiply realized.

[12] Smith and Medin (1981), p. 1.

[13] Langer (1979), p. 221.

[14] A similar criticism applies to DOF-based emergence. Every composite body will have less degrees of freedom than its components: that is just a way of stating what it means for a whole to made of parts.

Does it also apply to the concepts used by the special sciences? Fodor argues that they are different:

> ... there are special sciences not because of the nature of our epistemic relation to the world, but because of the way the world is put together: not all natural kinds (not all the classes of things and events about which there are important, counterfactual supporting generalizations to make) are, or correspond to, physical natural kinds.[15]

This is a strong assertion. It implies that the natural kinds picked out by the special sciences are mind-independent and give us direct insight into the structure of the world. I struggle with this idea for two reasons. The first is that the boundaries between special sciences are blurred. Chemistry blends into biochemistry which blends into molecular biology, cellular biology, neurology and neuroscience. Chemistry and biochemistry have direct implications for neuroscience. It's a tangle. Wikipedia lists around 60 branches of science beginning with the letter 'a'.

What's worse, the boundaries shift over time with new special sciences springing into being (cellular biology with the discovery of cells in the seventeenth century, genetics with the discovery of the gene in the 19th) and others fading away (phrenology or numerology). This process can happen in a matter of decades. Cognitive science developed in the latter half of the twentieth century. Yet in 2019 a paper[16] asked "What happened to cognitive science?" and answered that it had mostly been swallowed by other fields. If you've worked in academia, you're probably come across fierce battles over funding to decide exactly where in an overlapping spectrum of disciplines a project fits.

The idea of natural kinds neatly dividing up the world is an illusion. Does this lead to a postmodern world where anything goes? Although recent philosophical debate has not been kind to the idea of natural kinds, it doesn't go this far. A 2007 paper[17] by Ian Hacking concluded that some kinds may be more natural than others, but that there is no such thing as a natural kind. And an anthology from 2016 aims "to shift philosophical investigation of the naturalness of natural kinds to how they are used, discovered, or made".[18] We should treat "natural" kinds as those that help us answer the scientific

---

[15] Fodor (1974).
[16] Núñez et al. (2019).
[17] Hacking (2007).
[18] Kendig (2016).

questions we are interested in and accept that they change as our interests change.

This is a stark contrast to Fodor's assertion, which to me is an example of the mind projection fallacy (Sect. 4.5). Does it imply that entities with different cognitive makeup would have a different set of special sciences? I will return to this interesting question in Chap. 13.

Another response would be to say that whatever the reality, explanations in terms of concepts are better. Hilary Putnam takes the example of explaining why a square peg will not pass through a round hole. He asserts that the special science explanation which focuses on the macroscopic shapes of hole and peg is the best explanation because it picks out the relevant factors and ignores the irrelevant ones. By contrast, if a full microphysical explanation was available:

> I think that in terms of the purposes for which we use the notion of explanation, it is not an explanation. If you want to, let us say that the deduction is an explanation, it is just a terrible explanation, and why look for terrible explanations when good ones are available.[19]

It's hard to disagree that special science explanations are necessary for our imaginative understanding. But they are necessary because of our cognitive limitations, they are dumbed-down versions of the full explanation. Studying the concepts they use, and the multiple realization they entail, tells us much about how our brains work. But little about the structure of the world.

8.7 Approximations

Multiple realization is a consequence of conceptual thought. We cannot think without lumping things into categories. Such lumping must involve approximations. No two distinct things are identical (this is Leibniz's Principle of the Identity of indiscernibles). So to treat two distinct things as identical we need to abstract away from some of their differences. Mapping a 6 × 6 grid onto a 2 × 2 grid, as illustrated by Fig. 8.1, is one example of this. This section discusses the sort of approximations we need to make in real systems.

Let's start with an example from statistical mechanics. Our humble one litre box of gas is constantly shifting between a massive number of microstates. Imagine the richness and poetry of each of these microstates if only we had the ability to observe them. Imagine our senses were exquisitely

---

[19] Putnam (1975).

fine at both spatial and temporal scales. Imagine the intricate ballet of collisions, the endless slow-motion dance of molecule with molecule, the long periods where molecules assemble into geometric patterns before fading back into randomness.[20] In the whole history of the universe, no two microstates of the gas will ever be the same. And we describe all this it by just 3 variables!

Let's now take two "identical" boxes of gas. At any instant, they will have different microstates. Despite their being distinct at a microscopic level, we treat them as identical because we can describe them by the same three variables. But this, too, is an approximation. The constantly changing microstate causes the temperature of the gas to fluctuate. At room temperature, such fluctuations will be of the order of 10^{-10}K.[21] So even in terms of the aggregate variables, the boxes can only be similar, not identical.

Box 8.1: An emergent law of nature?

The Second Law of Thermodynamics states that, in an isolated system, entropy can never decrease.[22] It is often claimed to be an emergent law of nature due to two properties. The first is that it only holds in systems with large numbers of components, so "more is different". The second is that it is multiply realized in the sense that it holds independent of the details of the components of the system. It holds for gases, liquids, patterns such as the grid in Fig. 8.1 and pretty much everything else.

The Second Law is statistical in nature, saying no more than that improbable states will rarely be observed. This means that it is approximate. In any system, there will be constant short-lived violations of the law as random fluctuations take the system into lower entropy configurations. In systems at human scale, these fluctuations will be negligible. But in smaller systems they can be measured. A 2002 paper[23] reported the results of an experiment confirming their existence in a system comprising of latex beads with a diameter of 0.06 mm suspended in water. In general, the approximate nature of the second law means the size of violations depends in a well-defined way on the size of the system. And the precise dynamics of the processes that lead to violations will depend on the properties of the system's components. Each physical system is distinct. Multiple realization is no more than a useful approximation.

---

[20] This is of course a classical description. But that only reinforces my point about cognitive limitations. I don't know how to describe what a quantum microstate would look like.

[21] The calculation is in Sect. 7.3.

[22] For more details, see Graham (2023), Chap. 3.

[23] Wang et al. (2002).

Next, phase transitions and the example of boiling water. The theory treats them as discontinuous jumps (more about this in Chap. 10), with water at 100 °C minus a tiny amount being fundamentally different from water at 100 °C plus a tiny amount. But, if we look at a small range around boiling point, we find a hugely complex, but continuous, range of behaviour. To give you an idea what is involved, Fig. 8.3 is taken from a 2023 paper that reports the results of a nanoscale simulation of the dynamics of boiling water.

The images are simulations of the process of nucleation, the formation of bubbles, on surfaces. The two columns represent surfaces with different degrees of wettability or stickiness. Time increases from the bottom to the top. What they show are some steps in the complex, continuous process involved in a phase transition.

So much for water. Different liquids will behave differently depending on the exact nature of the bonding between their component molecules. Many details of the environment will affect this too: gravity; the nature of the surface; the presence of other substances. We neatly abstract away from all this when we apply the concept of boiling.

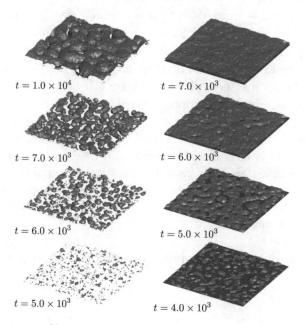

$t = 1.0 \times 10^4$ $t = 7.0 \times 10^3$

$t = 7.0 \times 10^3$ $t = 6.0 \times 10^3$

$t = 6.0 \times 10^3$ $t = 5.0 \times 10^3$

$t = 5.0 \times 10^3$ $t = 4.0 \times 10^3$

Fig. 8.3 Boiling water[24]

Now let's turn to universality. Take another look at Fig. 8.2. The author of the paper it is taken from notes: "…it will be seen that except for carbon monoxide and methane most of the points lie on or near a single curve".[25] This sounds like a description of the results of any experiment: the results are spread around the theoretical line, and the spread is narrower for some cases than for others. But it suggests that universality is only approximate. A 2009 survey[26] of empirical work in the area concludes that the best estimate for the critical exponent is $\beta = 0.326 \pm 0.002$. This precision is quite extraordinary given the formidable challenges involved in such experiments. But it is a long way from demonstrating strict universality. Chapter 10 gives a theoretical estimate for this precision. To get a ballpark idea, remember the statistical fluctuations in temperature I described in Sect. 8.2. These are no less than 6 orders of magnitude smaller than the error in the empirical values of the critical exponents. For the Ising model, empirical work[27] from 2020 finds a range from 0.322 to 0.372. It seems that universality is just another approximation, albeit a fascinating and useful one.

Every system is unique, but it is useful to treat them as if they were not. This response to multiple realization is known as *kind splitting*. The earliest reference I can find is from 1958.[28] It was widely deployed in response to Putnam's original arguments (Sect. 8.1) by saying that his concept of pain should be separated into reptilian pain, octopoid pain or human pain with distinct neurological properties. I don't see any way to escape it. Natural kinds cannot easily be resurrected. All that's left are half-hearted appeals to utility or common sense. Here's philosopher Daniel Stoljar:

> …as such there is no sense to be made in any attempt to theorize about pain as such—there is no such thing. But is this really true? Isn't it the case that we can generalize across species in certain circumstances? To put it another way, even if there is such a property as pain-in-humans, and even if it is appropriate to focus on it for certain theoretical purposes, is there not also such a property as being in pain simpliciter?[29]

Similarly, at the end of a discussion of a similar objection, Wilson concludes

---

[25] Guggenheim (1945).

[26] Sengers and Shanks (2009).

[27] Li et al. (2020), Table 1.

[28] Feigl (1958), reprinted as Feigl (1967), the argument is on p. 24.

[29] Stoljar (2010), p. 121.

…we have not been given good reason to resist taking the prima facie appearances of higher level reality seriously,[30]

To my ears, these responses have more than a touch of desperation. I find it particularly baffling that philosophers of mind are so intent on claiming multiple realization of brain states. For me, a cognitive concept like pain is simply a useful high level synopsis of a hugely complex underlying system. What does it even mean to say that the ache of eyes tired from reading that I feel now is that same as it was a year ago, or indeed a millisecond ago. Let alone the same across individuals. At best it implies that the states share some common features. That they are approximately the same, in precisely the same way that two boxes of gas are approximately the same.

8.8   The Role of Physics

We cannot avoid thinking in concepts. But I hope the previous section has illustrated how physics can help us understand the approximations involved. Let's return to the grids of Fig. 8.1 but this time with a tweak the rule than maps from the underlying 3 × 3 grid to a single element in the 2 × 2 grid. Instead of the rule being based on a majority, use the rule that the single cell is on or off depending on whether an even or odd number of the nine underlying cells are odd. Then let's imagine we can give the underlying 6 × 6 systems knocks, either small which just flip one element or large which flip more than one element.

Now let's observe the effects of the knocks as special scientists i.e. we only see the effect on the coarse grained 2 × 2 grid. What would we see? Small shocks (that only flip one cell) always change the state of the system. Large shocks (which flip more than one) change the state only 50% of the time. It's not hard to imagine baroque special science explanations, perhaps involving the term emergent fragility to explain this greater sensitivity to the small than to the large. But we can only understand it in a precise way, and one that will generalise to larger grids, when we can look under the hood and see the approximations involved in the "emergence" of the 2 × 2 grid.

Far from multiple realization preventing reduction, physics is the only way to understand why there is multiple realization. Chapter 10 introduces models of universal behaviour and shows how they explain what microphysical features systems must share to belong to the same universality class. Then it is a matter for physics to explain how these microphysical features arise.

---

[30] Wilson (2021), p. 84.

This is one reductionist explanation of how multiple realization works. For another, let's return to the paper[31] by Israeli and Goldenfeld I mentioned in the last chapter. This studies coarse graining in the Game of Life. They find that for a given grid, many coarse graining transformations are possible. Using the language of this chapter, for a given system can be described by many multiply realized high level concepts. It would be interesting to see this approach applied to scientific models.[32]

Next, let's think how the concept of temperature is multiply realized across systems. The physical state corresponding to a particular temperature is different in a solid, a liquid, a gas, a plasma or a radiating body. Yet the study of statistical mechanics leads to the zeroth law of thermodynamics which, based off the concept of thermal equilibrium, explains why the temperature of any two systems can be compared no matter what their underlying structure.

Physics can also go some way to explain why we make these particular approximations. It is mainly a function of our scale. Observations at a human scale are inevitably massively coarse grained in both space and time. We cannot avoid averaging over vast numbers of particles on time scales which are huge compared to the time scales associated with these particles. So then the question becomes why are we so large? The answer relates to thermal noise. In my previous book,[33] I show that such noise underlies everything that goes on inside cells (in fact, pretty much everything that goes on everywhere). But a cell has to be large enough so that the effect of this noise averages out to give some degree of stability. Otherwise, the structured processes that constitute a cell would be swamped by randomness. So thermodynamics implies a minimum size for a cell and hence a minimum size for aggregations of cells such as you or I. And this minimum size will be orders of magnitude larger than that of atoms, meaning that atoms seem small to us.

We can keep asking why. Averaging is only possible because microscopic quantities have specific statistical properties. And there are many high level regularities that don't involve averaging. Why is the universe like this? It's not hard to think of systems where this isn't the case. Take a house of cards. The structure of the whole depends on the precise position and properties of each individual card. Slightly change the angle or the mass of one card

---

[31] Israeli and Goldenfeld (2006).

[32] Rosas et al. (2024) investigates the nature and degree of autonomy of coarse graining in a range of systems. The motivating analogy is that of software that runs independently of the precise configuration of the electrons in the computer it runs on. However, this is just a restatement of the multiple realization argument and is subject to the same criticism. Independence is only approximate. Move a few electrons, a transistor flips state and the program crashes.

[33] Graham (2023).

and the house will collapse. Would a universe without high level regularities be equally fragile and so unable to support complex structures? This brings the issue into the domain of fine tuning arguments which maintain that basic properties of the world seem chosen to produce conditions that could support life. Fascinating as such questions are, they have nothing to do with emergence.

8.9   Discussion

Multiple realization seems an inevitable consequence of our being macroscopic creatures who make macroscopic observations. We neither have the sensory ability nor the processing power to follow every detail of what's going on, and these leads to our working with approximate concepts. Colin Klein writes that

> We cannot simply read the causal structure of the world just by looking at what nouns scientists use.[34]

It's hard not see that the much of the literature on multiple realization is suffering from the mind projection fallacy. The human mind does not play a role in structuring the physical universe. Nor do the special sciences. To return to Fodor's point, of course the concept of mountains will continue to play a useful role in geology. But it is strange to conflate the utility of a concept to humans with the nature of the universe.

8.10   Further Reading

There is a whole book on multiple realization, Polger and Shapiro (2016). Interesting attempts to use information theory to explain the structure of the special sciences are in Rosas et al. (2020) and Varley and Hoel (2022). For the sheer difficulty of measuring critical exponents, see Moldover et al. (1979) and Sengers and Shanks (2009).

A comprehensive discussion of the nature and role of concepts is in Murphy (2004). For more on natural kinds, see Bird and Tobin (2024). Heil (2003) has an interesting discussion of the relation between language and levels. Strevens (forthcoming) gives a philosophical account of why high

---

[34] Klein (2008).

level models are possible. A clear presentation of Putnam's argument about different sorts of pain is in Bickle (2020), Sect. 2.2.

More suggestions for reading can be found at www.TheMaterialWorld.net.

References

Bickle J (2020) Multiple Realizability. The Stanford Encyclopedia of Philosophy. https://plato.stanford.edu/entries/multiple-realizability/

Bird A, Tobin E (2024) Natural Kinds. The Stanford Encyclopedia of Philosophy. https://plato.stanford.edu/entries/natural-kinds/

Feigl H (1958) The "Mental" and the "Physical." In: Feigl H, Scriven M, Maxwell G (eds) Minnesota Studies in the Philosophy of Science. Volume II: Concepts, Theories, and the Mind-Body Problem. University of Minnesota Press, Minneapolis, pp 370–497

Feigl H (1967) The "Mental" and the "Physical": The Essay and a Postscript. University of Minnesota Press, Minneapolis.

Fodor JA (1974) Special sciences (or: The disunity of science as a working hypothesis). Synthese 28:97–115. https://doi.org/10.1007/BF00485230

Fodor JA (1997) Special Sciences: Still Autonomous After All These Years. Noûs 31:149–163. https://doi.org/10.1111/0029-4624.31.s11.7

Gallo M, Magaletti F, Georgoulas A, et al (2023) A nanoscale view of the origin of boiling and its dynamics. Nat Commun 14:6428. https://doi.org/10.1038/s41467-023-41959-3

Graham L (2023) Molecular storms: the physics of stars, cells and the origin of life. Springer, Cham, Switzerland. https://doi.org/10.1007/978-3-031-38681-7

Guggenheim EA (1945) The Principle of Corresponding States. The Journal of Chemical Physics 13:253–261. https://doi.org/10.1063/1.1724033

Hacking I (2007) Natural Kinds: Rosy Dawn, Scholastic Twilight. Roy Inst Philos Suppl 61:203–239. https://doi.org/10.1017/S1358246100009802

Heil J (2003) Levels of Reality. Ratio 16:205–221. https://doi.org/10.1111/1467-9329.00218

Israeli N, Goldenfeld N (2006) Coarse-graining of cellular automata, emergence, and the predictability of complex systems. Phys Rev E 73:026203. https://doi.org/10.1103/PhysRevE.73.026203

Kendig C (2016) Natural kinds and classification in scientific practice. Routledge, Abingdon

Klein C (2008) An ideal solution to disputes about multiply realized kinds. Philos Stud 140:161–177. https://doi.org/10.1007/s11098-007-9135-7

Langer SK (1979) Philosophy in a new key: a study in the symbolism of reason, rite and art. Harvard University Press, Cambridge (Mass.)

Li Z, Xia W, Su H, et al (2020) Magnetic critical behavior of the van der Waals Fe5GeTe2 crystal with near room temperature ferromagnetism. Sci Rep 10:15345. https://doi.org/10.1038/s41598-020-72203-3

Moldover MR, Sengers JV, Gammon RW, Hocken RJ (1979) Gravity effects in fluids near the gas-liquid critical point. Rev Mod Phys 51:79–99. https://doi.org/10.1103/RevModPhys.51.79

Morrison M (2014) Why Is More Different? In: Why more is different. Springer, New York

Murphy GL (2004) The big book of concepts. MIT Press, Cambridge, Mass.

Núñez R, Allen M, Gao R, et al (2019) What happened to cognitive science? Nat Hum Behav 3:782–791. https://doi.org/10.1038/s41562-019-0626-2

Polger TW, Shapiro LA (2016) The multiple realization book. Oxford University Press, Oxford

Putnam H (1967) Psychological predicates. In: Capitan WH, Merrill DD (eds) Art, Mind, and Religion. University of Pittsburgh Press., pp 37–48

Putnam H (1975) Philosophy and our mental life. In: Mind, Language and Reality Philosophical Papers, Volume 2. Cambridge University Press

Rosas FE, Geiger BC, Luppi AI, et al (2024) Software in the natural world: A computational approach to hierarchical emergence. https://doi.org/10.48550/arXiv.2402.09090

Rosas FE, Mediano PAM, Jensen HJ, et al (2020) Reconciling emergences: An information-theoretic approach to identify causal emergence in multivariate data. PLoS Comput Biol 16:e1008289. https://doi.org/10.1371/journal.pcbi.1008289

Sengers JV, Shanks JG (2009) Experimental Critical-Exponent Values for Fluids. J Stat Phys 137:857–877. https://doi.org/10.1007/s10955-009-9840-z

Smith EE, Medin DL (1981) Categories and concepts. Harvard University Press, Cambridge, Mass

Stanley HE (1987) Introduction to phase transitions and critical phenomena. Oxford University Press, New York Oxford

Stoljar D (2010) Physicalism. Routledge, London; New York

Strevens M (forthcoming) Why High-Level Explanations Exist. In: Wilson A, Robertson K (eds) Levels of Explanation. Oxford University Press

Varley TF, Hoel E (2022) Emergence as the conversion of information: a unifying theory. Philosophical Transactions of the Royal Society A: Mathematical, Physical and Engineering Sciences 380:20210150. https://doi.org/10.1098/rsta.2021.0150

Wang GM, Sevick EM, Mittag E, et al (2002) Experimental Demonstration of Violations of the Second Law of Thermodynamics for Small Systems and Short Time Scales. Phys Rev Lett 89:050601. https://doi.org/10.1103/PhysRevLett.89.050601

Wilson JM (2010) Non-reductive Physicalism and Degrees of Freedom. The British Journal for the Philosophy of Science 61:279–311. https://doi.org/10.1093/bjps/axp040

Wilson JM (2021) Metaphysical Emergence. Oxford University Press, Oxford

9

Weak Emergence: It's the Context

Summary Contextual emergence is the idea that the environment, or context, of a system affects its behaviour by imposing constraints on the underlying physics. This means that knowing the microphysics is not enough to explain the system; you need to know the context too. But contextual emergence fails as a useful definition. The split between system and context is entirely arbitrary. Further, the context is a physical system too and all contextual emergence ends up saying is that different physical systems interact. This is illustrated with examples including the evolution of the universe, feedback and selection in chemical systems and chemical computation.

Take an atom of silicon. Its behaviour will be different according to its environment. It might be part of a terrestrial rock, a microprocessor or a bone of my body. It might be in interstellar space and about to fall into a black hole. In each case it will behave differently but whatever the behaviour, it will always be consistent with physics.

This is an example of something general. Physics only puts broad limits on how a system can behave. The system's context determines which of the possibilities allowed by physics is realized. Contextual emergence says that this means reducing behaviour to low level physics is not possible, since the context is at a higher level. The term was introduced by Robert Bishop and Harald Atmanspacher in a 2006 paper.[1] In the last decade, it has been the

---

[1] Bishop and Atmanspacher (2006). The 1974 paper which introduced the term downward causation defines it in a way which sounds like contextual emergence: "I advocate not the autonomy of

© The Author(s), under exclusive license to Springer Nature
Switzerland AG 2025
L. Graham, *Physics Fixes All the Facts*, The Frontiers Collection,
https://doi.org/10.1007/978-3-031-69288-8_9

subject of three monographs, by Bishop, George Ellis, Michael Silberstein and Mark Pexton.[2]

The chapter starts by explaining the idea of contextual emergence. Once again, it turns out that the definition applies to everything: there are no physical systems for which we can ignore the context. In any case, the context does not somehow magically appear ex nihilo, but is a physical system too, and the split between system and context arbitrary. So all contextual emergence says is that different physical systems interact. There is no challenge to physicalism since interactions have always been the bread and butter of physics. I will illustrate this with examples including the evolution of the universe, feedback and selection in chemical systems and chemical computation.

Before starting, a general point. A fair part of the discussion of contextual emergence is about the way systems are modelled. In this chapter, I draw my examples from physical systems and will mostly defer a discussion of the role of models until the next chapter.

9.1   Contextual Emergence

The basic idea of contextual emergence is straightforward. While physics sets the necessary conditions for systems, it does not determine the sufficient conditions. A configuration of a system cannot contradict physics, but this is a weak criterion: there will generally be many possible configurations that are consistent with physics. This means physical causal closure is close to irrelevant. Think of that atom of silicon. Physics allows it to behave in a myriad of ways. What determines which one is realised? The system the atom is embedded in, its context: the rock, the microprocessor, the bone or the black hole.

The context manifests itself as a set of constraints. These constraints are outside the system so are not included in the physical description of the system. Hence they cannot be reduced to physics. This is contextual emergence. There is downward causation but it is benign. The constraints guide the underlying physical processes without violating physical causal closure.

---

higher levels, but rather the additional restraints, aspects of selective systems that these higher levels encounter." Campbell (1974), p. 182.

[2] Ellis (2016), Bishop (2019) and Bishop et al. (2022).

Here is a straightforward example of water flowing in a tap:

…the domain of elementary particles contributes some of the necessary conditions for the existence of the properties and behaviors of water flowing through a faucet: no elementary particles and forces, no flowing water. Nonetheless, the existence of elementary particles and their forces do not guarantee that fluids flowing out of faucets will exist. The total set of necessary and sufficient conditions for such flowing fluids is, itself, contingent rather than necessary, and involves more conditions than are found in the domain of elementary particles and forces.[3]

Contextual emergence captures the distinction between a system and its environment. In the example, the properties of the system, the flowing water, are constrained by their environment of the tap. Subject to these constraints, the water follows the laws of physics. But the constraints themselves are to be found nowhere in the physics of water. Examples of constraints include

…conservation laws, free energy principles, least action principles, symmetries, and some types of symmetry breaking. More specifically, think of the light postulate and the relativity principle as instances of adynamical global constraints—they delimit the kind of dynamics physically accessible at all spatial and temporal scales.[4]

Constraints can manifest themselves as stability conditions. A particularly fascinating example of a stability condition is the dimensionality of space. It is simply assumed in the underlying physics but affects everything. For example, stable planetary orbits do not exist in spaces with more than three dimensions.[5]

The work discussing contextual emergence includes a host of case studies. Here are two. Lasers only exist because of the context of an optical cavity containing a specific mix of atoms and the triggering of a particular configuration of those atoms.[6] Or the text on this page which, while nothing but an arrangement of atoms, only has meaning for you because of a whole host of contexts including the rules of syntax and the intentions of all the people involved in the publishing process.

---

[3] Bishop (2019), p. 3.3.
[4] Bishop et al. (2022), p. 28.
[5] See Barrow (1983).
[6] Both these examples are taken from Bishop (2019).

Box 9.1 Synchronic and diachronic emergence

Multiple realization and contextual emergence involve the distinction between whole and parts at just one point in time. This is synchronic or cotemporal emergence. Diachronic emergence, on the other hand, says that the relevant distinction is between the system at two different times. This is the case for many self-organising systems where their complexity increases during their evolution. It is implicit in the idea of simulation emergence which is concerned with tracking properties over time.

In a recent paper, Wilson argues convincingly that cases of diachronic emergence can be seen either as cases of synchronic emergence or turn out to be instances of causation between levels. She concludes that "there isn't any need for a diachronic notion of metaphysical emergence".[7]

The distinction between synchronic processes happening at a point in time and diachronic processes happening over time will resurface at various points in what follows, particularly in the discussion of the different between understanding and predictability in Chap. 14.

Some authors describe hierarchies of contexts, with the contexts at higher levels determining those at lower ones. Ellis describes five levels of context giving rise to five types of downward causation[8]: deterministic downward causation (this is contextual emergence as discussed so far), non-adaptive information control (feedback loops allowing the attainment of goals), adaptive selection (for example, evolution); adaptive information control (feedback guided by adaptive selection); adaptive selection of selection criteria (the result of one process of adaptive selection guides another such process). In a book fascinating and frustrating in equal measure, but sadly devoid of concrete examples, Terrence Deacon describes multiple levels of contextually emergent processes.[9] These are illustrated in Fig. 9.1. At the lowest level are homeodynamic (effectively thermodynamic) processes. The give rise to complex structures (morphodynamics) then to intention (teleodynamics) and ultimately consciousness.

Deacon's description makes clear the link to contextual emergence. Necessary conditions come from the next-lowest level, shown by the upward arrows in the diagram. Sufficient conditions come from the next highest level, shown by the downward arrows. A further link is to self-organisation (which happens in the middle level of Deacon's hierarchy, morphodynamics). The local behaviour of an individual in a flock only makes sense in the context of

---

[7] Wilson (Forthcoming).

[8] Ellis (2016), Chap. 4.

[9] Deacon (2012).

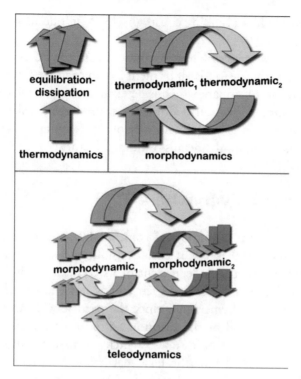

Fig. 9.1 Three levels of emergent dynamics[10]

the whole flock, local dynamics are subject to global dynamics, the individual is subject to the context.

There is also a close relation between contextual emergence and the ideas discussed in the previous chapter. In contextual emergence, physics sets the necessary but not the sufficient conditions. In MR emergence, physics sets the sufficient but not the necessary conditions since many low level systems can produce the same high level one. We saw that the definition of DOF emergence nests MR emergence. It also nests contextual emergence. To see this, take a system comprised of two stationary particles. If these are free there are 6 degrees of freedom (the 3 coordinates of each particle). If they are rigidly fixed together there are only 5 degrees of freedom since they are constrained to be a fixed distance apart. Then let's impose a contextual constraint by joining the two particles by a flexible but inelastic string. We still have 6 degrees of freedom, but the range of them is limited by the length of the string. In

[10] Source: Deacon (2006), Fig. 5.8 © Oxford University Press 2006. Reproduced with permission of the Licensor through PLSClear.

general, if a system hits a binding constraint, by definition its must lose some degrees of freedom.

Which of the examples of Chap. 2 can be described as contextually emergent? Once again, the answer is all of them. As an example from the smallest scale, the behaviour of a quark which is determined by the nucleonic context in which it is found. Then an example from the largest scale: the relevant context for the atom in Churchill's nose involves pretty much everything else in the universe, including the details of human history.

9.2   Context is Everywhere

The first issue with the definition of contextual emergence is the same as raised in the previous chapters: it excludes nothing but the most idealised of systems. What system is independent of its context? An isolated particle in an empty universe, maybe. But even that is not enough of an idealisation to avoid contextual emergence since the particle will interact with the virtual particles produced by the vacuum. So to find a system which is not contextually emergence we need to also switch off these quantum effects.

In Sect. 6.6, I showed how the idea that "more is different" is implicit in the Hamiltonian description of a physical system. Exactly the same argument applies to contextual emergence. In the Hamiltonian the energy of one particle depends, in general, on the position and momentum of all the others. We can always make an arbitrary choice to limit our modelling to a subset of the particles and ignore the rest. We then call the subset the experiment, the rest the context and live with the approximation involved. Such a choice will depend on the question being addressed.

Take the example of an atom in the earth's crust. To fully understand its motion, you need the whole universe and its history. However if you are just interested in how it vibrates in its crystal, the context becomes the lattice or perhaps just a few surrounding atoms, along with the temperature and pressure. If you are interested in the movement of the atom through the crust, the context is the large-scale structures of which it is part. If you are interested in the movement of the atom through space, the context is the planet and the solar system. And so on. The distinction between system and context is a function of our interests and computational resources.

Lurking here is the same problem as I raised in the previous chapter: you cannot read off the nature of reality from what scientists do. Yet this is a key part of contextual emergence. Bishop et al. write that:

Our argument will be that the plethora of explanatory pluralism in the sciences turns out to be good evidence for contextual emergence.[11]

I disagree. Scientific practice mostly tells us about human capabilities, not about the structure of the universe.

9.3   The Context is Physics

Contextual emergence treats the context as given. Yet the context is a physical system too. It may be an arbitrary choice which part of the universe is defined as the system and which part is the context. But wherever the division lies, both are physical systems. This means the causation is not from the context, treated as given, to the system, but from the microphysical level of the context to the microphysical level of the system. Everything is physics.

To investigate this, let me take two examples used by proponents of contextual emergence. The first of these is a process that occurs in plasmas called Debye screening. The electric field of one electron (the system) is influenced by all those around it (the context) and varies substantially from the field of an isolated electron. This is shown in Fig. 9.2 which shows the electron's electric potential on the y-axis against distance on the x-axis. The top line is the standard exponential potential of an isolated electron. The screened potential that arises from the interaction is the lower line.

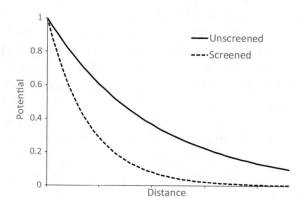

Fig. 9.2 Debye screening

---

[11] Bishop et al. (2022), p. 23.

In terms of contextual emergence, this means that:

>...an electron is no longer the same bundle of properties in a complex aggregate environment as it is in isolation.[12]

This interpretation is reinforced by the *effective field* approach which is used to solve the system. This involves calculating the potential for one electron, the system, while taking that of all the others as given, the context. I'll have more to say about this in the next chapter, but it seems to lead naturally to the idea of contextual emergence.

Except that such a modelling approach is just a useful approximation. The world knows nothing of effective theories. What is actually going on is that the many electrons in the system are interacting in complicated ways. Such a problem might seem intractable, but recently two neat papers[13] have studied Debye screening as a many-body problem. They describe the mechanism behind it in terms of electrons deflecting off one another. In principle, you could use their results to start with a single electron and show how the screening effect increases as you progressively add electrons. Instead of taking the context as given, these papers show how the context is just another physical system. The trouble is, of course, that solutions to many-body problems are rare. But the fact that we can't as yet solve a problem doesn't mean that a phenomenon is caused by anything beyond the complicated interaction of many bodies.

What about stability conditions? The dimensionality of space has a fundamental role in what structures are possible and it is clearly outside quantum physics. But it is no more a candidate for a driver of emergence than are the values of the fundamental constants, none of which are explained by physics either. We can again wrap this up with the debate on fine-tuning. Yes, if the constants took different values, the world would be different. Maybe some future physics will explain why the constants take the values they do. Or maybe not. Whatever the case, what has this got to do with emergence?

While some stability conditions are issues of fine tuning, others are the result of physics. The best-known example is the problem of the stability of matter. Ordinary matter is made up of electrons with negative charge and nuclei with positive charge. A knowledge of electromagnetism might lead you to expect electrons to fall into nuclei. One of the early successes of quantum mechanics was to explain why this doesn't happen and why ordinary matter is stable.

---

[12] Bishop et al. (2022), p. 170.
[13] Escande et al. (2015) and Escande et al. (2018).

A further sense of the stability of matter is this: if you double the number of atoms, do you also double their energy and the volume they occupy? Without this exact doubling, ordinary matter could not exist. In the late 1960s, the first proof of this was given by Freeman Dyson and Andrew Lenard.[14] The proof has the fascinating consequence that stability does not hold for a system comprised of just bosons. Stability turns out to be a consequence on the Pauli exclusion principle so requires that either the positive or the negative charges are fermions.

Turning now to thermodynamic stability, standard derivations can be found in many textbooks.[15] The basic equilibrium concept in thermodynamics is that of temperature, via the zeroth law, and temperature can be defined from quantum physics.[16] Then onto conservation laws. These are not magically part of the context but the consequence, via Noether's theorem, of symmetries of the universe. Of course, we can ask why we live in a universe in which Noether's theorem holds, or why it has the symmetries it does, but these are yet more fine-tuning questions and have nothing to do with emergence.

Other stability conditions claimed as examples of contextual emergence turn out to be properties of models, not the world. Chap. 15 will discuss one of these, the Born–Oppenheimer approximation used in calculations of molecular structure.

Stability conditions are certainly interesting. But they are either outside current physics in a fine tuning sense. Or they can be explained by physics. In neither case do they play the magical role assigned to them by proponents of contextual emergence.

An important part of the motivation for contextual emergence seems to be a sense of astonishment that the rich complexity of the world somehow evolved from conditions in the early universe. Here's Bishop:

> … how would the initial conditions plus the laws yield all of the contingent conditions leading to the emergence of macroscopic systems, the evolution of galaxies, stars and solar systems, life on planet Earth, the contingent species types and numbers of organisms, and so forth?[17]

And here's Ellis again, arguing that a reductionist view implies

---

[14] Dyson and Lenard (1967) and Lenard and Dyson (1968).
[15] For example Waldram (1985), Chap. 3.
[16] Gemmer et al. (2004), Chap. 3.
[17] Bishop (2019), p. 5.8.

....that the particles existing when the cosmic background radiation was decoupling from matter, in the early Universe, were placed precisely so as to make it inevitable that 14 billion years later human beings would exist, Charles Townes would conceive of the laser , and Edward Witten would develop string theory... [instead] Top-down causation takes place as well as bottom-up action, with higher level contexts determining the outcome of lower level functioning, and even modifying the nature of lower level constituents.[18]

I'll respond to this in two ways. Firstly, with a rhetorical question of my own: if the context isn't a physical system and hence encoded in the early universe, where did it come from?

A rigorous response is based on information. The time evolution of quantum states is described by unitary operators. A property of such operators is that they can be applied either backwards or forwards in time so the state of a system at one point in time is sufficient to recover the system at any other point in time.[19] Thus the total information in an isolated system must be constant. Leonard Susskind calls this the minus first law "...because it underlies everything else. It says that information is never lost".[20]

Applying this to the universe as whole means that the informational content of the universe an instant after the big bang (when to our eyes it looked like a nearly uniform cloud of particles and radiation) is the same as the informational content of the universe now with all its rich structures (including us). And it is the same as the informational content when, in the distant future, the last black holes evaporate and the universe reaches a state of heat death (when to our eyes in would look like a nearly uniform cloud of radiation at a fraction about absolute zero).[21] Of course this sounds ridiculous. Our intuition completely fails to grasp it. So much the worse for intuition. Quantum physics gives us representational access to something beyond our imaginative understanding.

---

[18] Ellis (2005).

[19] Whether this can be done in practice is an interesting question. I will discuss it further in Sect. 14.5, along with the general distinction between understanding a system and being able to predict its evolution.

[20] Susskind and Friedman (2014), Sect. 4.1.

[21] I am of course glibly brushing over the black hole information paradox. See Maldacena (2020) for a review.

9.4   Example: Context in the Early Universe

An important part of the context for anything relating to what happens on our planet is the fact that it is part of a solar system. Here's the sketch of an explanation of the existence of solar systems from the conditions of the early universe[22]:

1. The inflation that followed the big bang had two important consequences:

    a. The distribution of matter and radiation in the early universe is almost, but not quite, uniform
    b. The temperature of the early universe falls sufficiently fast so that nucleosynthesis incomplete leaving a mix of around 75% hydrogen and 25% helium

2. The non-uniformities in the matter distribution lead to the cloud collapsing under gravity into cores. As the cores collapse, their temperature rises.
3. At some point in the collapse process, the temperature of the cores becomes high enough to start nuclear fusion.
4. Fusion offsets gravitational collapse. The resulting equilibrium is stable: the cores have become the first stars
5. When these stars have burnt most of their hydrogen, fusion can no longer offset gravity and the collapse process continues, catastrophically, to cause supernova.
6. The extreme conditions in supernova allow fusion to continue to heavier elements and the resulting explosion distributes them into the cloud.
7. The collapse process starts again and the next generation of stars are surrounded by clouds of heavier elements.
8. Collapse processes occur in these solar clouds, resulting in planets.

Of course, there are many open questions, though many of them are about conditions in the very early universe and examples of fine tuning. But still we have a story, based on nothing outside physics, which explains how the context of solar systems arises. The story can be continued to the origin of life (with, of course, many more open questions).

The interested reader should compare my account with that of Ellis.[23] Where I see physical processes, he sees context. Here's one example:

---

[22] A more thorough treatment is in Graham (2023), Chap. 6.
[23] Ellis (2016), pp. 6–7.

...the evolution of galaxies is a strong function of environment, because galaxies entering high density regions like clusters are prone to processes like ram pressure stripping, starvation, strangulation, and tidal stripping. These physical processes are significant because they influence the star formation rates and colours of galaxies as a function of environment.

The important point is that these physical processes are not magic but can be traced back to their origin in the big bang. They will arise naturally in a simulation of galaxy formation and I'll discuss one such simulation in the next chapter. The environment or context which Ellis describes is also physics.

9.5   Example: Feedback and Selection

Levels 2 and 3 of Ellis's hierarchy of downward causation are feedback control and adaptive selection. This section shows that these properties arise in simple chemical systems. These are exactly analogous to such properties in all chemical systems, including those that constitute life, the most interesting chemical system of all.

Negative feedback occurs when a system reacts to a change by tending to offset the change. It is a basic property of any chemical reaction.[24] Le Chatelier's principle states that chemical reactions tend to adjust to a change by counteracting the effect of the change. Take a chemical reaction which transforms reactants A into products B (in general the letters can represent more than one chemical)

$$A \rightleftharpoons B$$

If the reaction is in equilibrium the concentration of reactants and products is constant. Now if we add more of the reactants A, the reaction will proceed from left to right, reducing the concentration of A and increasing that of B until a new equilibrium has been reached. If we add more B, the reactions will proceed from right to left reducing the concentration of B. This is negative feedback.

Now let's introduce the idea of a catalyst. Catalysts are substances which increase the rate of a reaction. If a molecule C catalyses the reaction from A to B, we can write:

$$A + C \rightleftharpoons B + C$$

---

[24] This is a consequence of the principle of detailed balance. For more details, see Graham (2023), p. 91.

Note the catalyst appears on both sides, it participates in the reaction but is not changed by it. Catalysts can achieve dramatic rate increases, factors of the order of a billion are typical.

Catalysts are where the idea of selection enters chemistry. Let's imagine the reactant A can now undergo two separate reactions, only the first of which is catalysed:

$$A + C \rightleftharpoons B + C$$
$$A \rightleftharpoons D$$

Assume that the equilibrium of both reactions lies on the right. Further assume that in the absence of the catalyst both reactions happen so slowly we can ignore them. This means that a solution of A will be stable. Now add the catalyst C. This increases the rate of the first reaction so the product B will dominate. We can say the catalyst has selected B over D. This is known as kinetic selection.

Such kinetic selection is fundamental to the workings of living cells and also underlies all natural selection. To see this, let's take a slightly more complicated example. Some reactions are special in that the catalyst is also products of the reactions. Such reactions are called autocatalytic, they catalyse themselves. The simplest example of an autocatalytic reaction is:

$$A + C \rightleftharpoons 2C$$

The product C is also the catalyst. Let's compare this with an ordinary reaction, catalysed by a different catalyst C':

$$A + C' \rightleftharpoons B + C'$$

Assume again that the equilibrium of both reactions lies with the products but under normal conditions the reactants are stable. Further, let's assume that one molecule of the catalyst allows the respective reaction to happen a million times per second.

If we drop a molecule of C' into a beaker of A, how long will it take until we have a mole of B?[25] At a million reactions per second the answer is 6×10^{17} s, longer than the age of the universe. Now take the autocatalytic

---

[25] There are 6×10^{23} molecules in a mole. If a reaction doubles the number of molecules of a catalyst at each step, we need 79 steps to make a mole of the catalyst since $2^{79} \approx 6 \times 10^{23}$. If each reaction takes a millionth of a second, 79×10^{-6} is around 10^{-4} s. If each reaction takes 1,000 s, $79 \times 1,000 = 79,000$ s ≈ 22 h.

reaction. If we start with a beaker full of A and one molecule of C, then a millionth of a second later we have two molecules of C, another millionth of a second later four and so on. If this process continues, we will have a mole of C in around a ten-thousandth of a second. Even if the autocatalytic reaction were a billion times slower, producing one new molecule every 1000 s, it would still take a bit less than a day to produce a mole of C.

In practice A will quickly be used up locally and the reaction will stop until more A diffuses in from the environment. This will react quickly to produce more C and the process will go on until there is no more A. In more evocative language, the reaction transforms its environment from one dominated by the reactants to one dominated by the products. The reactants are driven to extinction. The remaining chemicals are adapted to reproduce rapidly in their environment. This starts to sound like natural selection and indeed some authors see such systems as central to the understanding of life:

> The fact that these capacities are typical of both inanimate replication and biological reproduction has far reaching consequences for the origin of animate matter.[26]

Bishop and coauthors contend that natural selection is contextually emergence and "…cannot be derived from any fundamental physics".[27] Yet natural selection is nothing but kinetic selection writ large. And much of chemistry can be derived from quantum physics either exactly or approximately.

Catalysed reactions also allow us to introduce a more sophisticated form of feedback control. Let's go back to the basic reaction

$$A + C \leftrightharpoons B + C$$

but with the added assumption that the presence of B reduces the catalytic effectiveness of C. So as B accumulates, the reaction slows down then stops. Then if another reaction starts using B, the reaction will start again until a new balance is reached. This is negative feedback and is precisely the mechanism which serves as a key regulator of metabolism in living cells.

The first step in the metabolism of glucose involves it being converted into another sugar by a reaction catalysed by an enzyme (a biological catalyst). The sugar can weakly bind to the enzyme and in doing so deactivates it. So the higher the concentration of the sugar, the less enzymes are available

---

[26] Lifson (1997).
[27] Bishop et al. (2022), p. 10.

to catalyse the reaction and the slower it proceeds. As the sugar is used in other reactions, its concentration falls, the enzymes reactivate and the reaction speeds up again. In more evocative terms, this mechanism allows the cell eats glucose until it is full up, then stop eating until it is hungry again. Enzymes that work in this way are called *allosteric* and control many cellular reactions.

All of this is straightforward chemistry. Context or emergence are nowhere to be found. So much for the first two types of downward causation in Ellis's hierarchy. The others are adaptive information control and adaptive selection of selection criteria. These are properties of much more complicated systems and I will briefly return to them in the final chapter.

9.6   Example: Chemical Computation

A common theme running through Ellis's work is that downward causation occurs in digital computers. In a paper with Barbara Drossel, the claim is that computational processes:

> … are examples showing that a simple materialist position—the idea that only physical entities can have causal powers—cannot be correct:

> Abstract entities have causal powers: It is clear from this discussion that (1) algorithms, (2) computer programs, and (3) data - all abstract entities – have causal powers because they alter physical outcomes in a real world social context.[28]

Just before this, the authors give cite a passage which includes this sentence:

> A computational process is indeed much like a sorcerer's idea of a spirit. It cannot be seen or touched. It is not composed of matter at all.[29]

As well as being incorrect, this neatly captures what I find objectionable about the way the concept of emergence is used. Let's see that there is nothing magical about computational processes and that they can happen with evoking sorcerers or, indeed, downward causation. As is often the case,

---

[28] Ellis and Drossel (2019).

[29] This quote comes from the introduction to a standard textbook on programming, Abelson et al. (1996), p. 2. The authors seem to me to intend it as metaphor rather than metaphysics.

evocation of magic is merely a sign that something is beyond our imaginative understanding.

To construct a computer, we need logic gates and a way of connecting them. This can be done in silicon and copper wire. Or it can be done using chemical reactions. Signals are represented by the concentration of a molecule, or of a particular state of a molecule. A high concentration represents a 1, a low concentration represents a 0.

The simplest example of a gate is a signal repeater which copies input into output. To see how this can be implemented, let molecule C be the input and A the output. A needs to exist in two different forms, an unexcited state A and an excited state A*. We can produce A* from A using a chemical P which acts as a power supply[30]:

$$A + P^* \leftrightharpoons A^* + P$$

A decay reaction tends to reduce the concentration of A* to zero.

$$A^* \leftrightharpoons A$$

Finally, assume that C catalyses the decay reaction so:

$$A^* + C \leftrightharpoons A + C$$

If the concentration of C is low, this reaction only proceeds slowly so the charging reaction dominates and the concentration of A is low. If the concentration of C is high, the decay reaction dominates and the concentration A of is high. In other words, the concentration of C, the input, is copied into the concentration of A, the output. This is a signal repeater or a YES gate.

All other logic gates can be constructed in similar ways. The details are fearsome and not particularly relevant here. A paper[31] from 1996 shows that such logic gates can be linked into arbitrarily large networks. This is all that is needed to show that a chemical computer can be Turing complete. There is much fascinating work in this area, both in constructing practical chemical computers and in using the insights to understand the goings-on in cells.

Now my argument is the same as in the previous section. Of course the behaviour of each reaction depends on the environment it is embedded in. But this environment is a designed network of chemical reactions and I can

---

[30] In fact, it must act like a molecular ratchet. For more details, see Graham (2023), Chap. 5.
[31] Magnasco (1997).

understand this from the bottom-up by following chemicals through their reaction paths (probably with the help of a simulation).

If you want, you can take an incremental approach, like the one I discussed in Sect. 9.3 in the context of Debye screening. Start with one reaction. Add in a second. Once you've thoroughly understood that, add in a third. Perhaps now you have a complete gate. Study its properties. Then add in another gate. And so on up to the full system. That will give you as complete a scientific understanding of the system as you have time for. Once more, the concept of emergence adds nothing, the context is just more chemistry.

Of course, Ellis's digital computer is built of transistors rather than chemical reactions. But if we want, we can unpick what happens in a computer just as we can unpick a chain of chemical reactions or the patterns in the Game of Life.

9.7   Discussion

The criticism offered by contextual emergence is directed against the sort of straw man I discussed in Chap. 6. None of it is the slightest challenge to reductive physicalism. The distinction between system and context is a reflection of our cognitive limitations. We are not smart enough to deal with everything at once so need to treat most of the universe as given, the context, and focus on one thing, the experiment.

I introduced contextual emergence with two examples. The first was the laser, which wouldn't exist without a lasing cavity, carefully constructed by humans. The second was the way in which atoms on this page translate into meaning in your brain. Indeed, a full understanding of both of these would require understanding everything in the universe. They are variations on the Churchill's Nose example of Sect. 2.15. But unless you think there was divine intervention somewhere along the way, every step that led up to them was caused by the interaction of physical processes which started with the big bang and gradually produced the physical context which makes lasers or meaning possible.

9.8   Further Reading

The reference for contextual emergence is Bishop et al. (2022); Bishop (2019) is also useful. For more details on Debye screening, see McComb (2007), Sect. 3.2, for the stability of matter see Lieb and Seiringer (2010) and for a review of the state of the art in many-body systems see Defenu et al. (2024).

Silva (2013) is a thorough overview of chemical computation and interesting examples can be found in Fages et al. (2017) and Dueñas-Díez and Pérez-Mercader (2019). For a detailed elucidation of the mechanism of a laser, see Kuhlmann (2014). This paper also includes an interesting discussion of Synergetics, a theoretical framework for self-organisation due to Hermann Haken, see Haken (1983). A recent review of fine tuning arguments is Hossenfelder (2021).

More suggestions for reading can be found at www.TheMaterialWorld.net.

References

Abelson H, Sussman GJ, Sussman J (1996) Structure and interpretation of computer programs. McGraw-Hill, Cambridge, Mass.

Barrow JD (1983) Dimensionality. Phil Trans R Soc Lond A 310:337–346. https://doi.org/10.1098/rsta.1983.0095

Bishop RC (2019) The physics of emergence. Morgan & Claypool Publishers, San Rafael, CA

Bishop RC, Atmanspacher H (2006) Contextual Emergence in the Description of Properties. Found Phys 36:1753–1777. https://doi.org/10.1007/s10701-006-9082-8

Bishop RC, Silberstein M, Pexton M (2022) Emergence in context: a treatise in twenty-first century natural philosophy. Oxford University Press, Oxford.

Campbell DT (1974) 'Downward Causation' in Hierarchically Organised Biological Systems. In: Ayala FJ, Dobzhansky T (eds) Studies in the Philosophy of Biology. Macmillan Education UK, London, pp 179–186

Deacon TW (2006) Emergence: The Hole at the Wheel's Hub. In: Clayton P, Davies PCW (eds) The re-emergence of emergence: the emergentist hypothesis from science to religion. Oxford University Press, Oxford; New York

Deacon TW (2012) Incomplete nature: how mind emerged from matter. W.W. Norton & Co, New York

Defenu N, Lerose A, Pappalardi S (2024) Out-of-equilibrium dynamics of quantum many-body systems with long-range interactions. Physics Reports 1074:1–92. https://doi.org/10.1016/j.physrep.2024.04.005

Dueñas-Díez M, Pérez-Mercader J (2019) How Chemistry Computes: Language Recognition by Non-Biochemical Chemical Automata. From Finite Automata to Turing Machines. iScience 19:514–526. https://doi.org/10.1016/j.isci.2019. 08.007

Dyson FJ, Lenard A (1967) Stability of Matter. I. Journal of Mathematical Physics 8:423–434. https://doi.org/10.1063/1.1705209

Ellis G (2016) How can physics underlie the mind? Springer, Berlin

Ellis G, Drossel B (2019) How Downwards Causation Occurs in Digital Computers. Found Phys 49:1253–1277. https://doi.org/10.1007/s10701-019-00307-6

Ellis G (2005) Physics, complexity and causality. Nature 435:743–743. https://doi.org/10.1038/435743a

Escande D, Bénisti D, Elskens Y, et al (2018) Basic microscopic plasma physics from N-body mechanics. https://doi.org/10.48550/arXiv.1805.11408

Escande DF, Elskens Y, Doveil F (2015) Direct path from microscopic mechanics to Debye shielding, Landau damping and wave-particle interaction. Plasma Phys Control Fusion 57:025017. https://doi.org/10.1088/0741-3335/57/2/025017

Fages F, Le Guludec G, Bournez O, Pouly A (2017) Strong Turing Completeness of Continuous Chemical Reaction Networks and Compilation of Mixed Analog-Digital Programs. In: Feret J, Koeppl H (eds) Computational Methods in Systems Biology. Springer International Publishing, Cham, pp 108–127

Gemmer J, Michel M, Mahler G (2004) Quantum thermodynamics: emergence of thermodynamic behavior within composite quantum systems. Springer, Berlin

Graham L (2023) Molecular storms: the physics of stars, cells and the origin of life. Springer, Cham, Switzerland. https://doi.org/10.1007/978-3-031-38681-7

Haken H (1983) Synergetics: an introduction: nonequilibrium phase transitions and self-organization in physics, chemistry, and biology. Springer, Berlin

Hossenfelder S (2021) Screams for explanation: finetuning and naturalness in the foundations of physics. Synthese 198:3727–3745. https://doi.org/10.1007/s11229-019-02377-5

Kuhlmann M (2014) A Mechanistic Reading of Quantum Laser Theory. In: Why more is different. Springer, New York

Lenard A, Dyson FJ (1968) Stability of Matter. II. Journal of Mathematical Physics 9:698–711. https://doi.org/10.1063/1.1664631

Lieb EH, Seiringer R (2010) The stability of matter in quantum mechanics. Cambridge University Press, Cambridge, UK

Lifson S (1997) On the Crucial Stages in the Origin of Animate Matter. J Mol Evol 44:1–8. https://doi.org/10.1007/PL00006115

Magnasco MO (1997) Chemical Kinetics is Turing Universal. Phys Rev Lett 78:1190–1193. https://doi.org/10.1103/PhysRevLett.78.1190

Maldacena J (2020) Black holes and quantum information. Nat Rev Phys 2:123–125. https://doi.org/10.1038/s42254-019-0146-z

McComb WD (2007) Renormalization methods: a guide for beginners. Oxford University Press, Oxford

Silva AP de (2013) Molecular logic-based computation. RSC Publ., Royal Soc. of Chemistry, Cambridge

Susskind L, Friedman A (2014) Quantum mechanics: the theoretical minimum. Basic Books, New York

Waldram JR (1985) The theory of thermodynamics. Cambridge University Press, Cambridge; New York

Wilson JR (Forthcoming) On the Notion of Diachronic Emergence. In: Bryant A, Yates D (eds) Rethinking Emergence. Oxford University Press, Oxford

10

Weak Emergence: The Map and the Territory

Summary Many definitions of emergence refer to the properties of models rather than properties of the world. In this chapter, I will argue that this is a mistake. Models use simplifications and abstractions to isolate what is common across different systems. They involve multiple realization, reduced degrees of freedom and the separation between system and environment; many can only be solved by simulation. By necessity, they fit the various definitions of weak emergence. But the map is not the territory, the properties of models are not the properties of the systems they represent. The chapter addresses this question by discussing three modelling strategies: the Thermodynamic Limit, Effective Field Theories and the Renormalization Group Transformation. These are illustrated with a detailed study of the Ising model.

The preceding chapters illustrated weak emergence with examples mostly drawn from real systems. This chapter turns to the role of models. Models are abstractions which make complex reality accessible to us. The best cut away all the extraneous detail to bring into focus what we are interested in. As such, models are always approximations and the best models often involve the most drastic approximations.[1] They isolate what systems have in common and often abstract from low level details.

---

[1] I am brushing over the distinction between approximation and idealization. Imagine describing a body falling through air. You could use an approximate model in which $v = at$. Or you could assume that the body is falling in a vacuum – an idealization – and use the exact model $v = at$.

© The Author(s), under exclusive license to Springer Nature
Switzerland AG 2025
L. Graham, *Physics Fixes All the Facts*, The Frontiers Collection,
https://doi.org/10.1007/978-3-031-69288-8_10

Models are central to all three types of weak emergence. For simulation emergence, the style of modelling, whether closed-form or simulation, constitutes the definition. Models naturally lead to the idea of multiple realization (MR) or degrees of freedom (DOF) emergence since usually one model can be used to describe many different systems. And the split between model and environment is at the heart of contextual emergence. There is even a definition of emergent properties as those which differ from a model.[2]

Given all this, it is unsurprising that weak emergentists tend to dwell on the properties of models. I think this is a mistake. Emergence, if it exists in a meaningful way, must be a property of physical systems. But the map is not the territory. The models we use are a function of our interests, our cognitive capabilities and the computational resources we have available. Extreme caution must be taken when trying to infer the properties of reality from the properties of models.

10.1   More Is the Same; Infinitely More Is Different

...the infinite is nowhere to be found in reality, no matter what experiences, observations, and knowledge are appealed to.[3]

Physicists like taking infinite limits.[4] It generally makes the maths easier since one can use calculus and, mostly, ignore boundary conditions. The thermodynamic limit is one such infinite limit. If you have a system consisting of N particles in a volume V, taking the thermodynamic limit means that you let both go to infinity in such a way the ratio N/V stays constant. In the limit the size of fluctuations goes to zero so they can be ignored and many useful quantities like entropy are independent of the size of the system.[5] Taking the limit needs to be done with care. It may not exist, for example in the case of gravitational systems. Or it can give results which contradict conservation laws.[6] But it is central to most statistical mechanics.

Let's take the example of phase transitions. Here's a description of boiling water:

---

[2] Cariani (1991).
[3] Hilbert (1998), p. 191.
[4] I borrow the title of this section from Kadanoff (2009).
[5] For a formal statement, see Appendix A.2.
[6] Lanford (1975).

When the temperature hits 100 degrees Celsius, the liquid water does something remarkable. It begins to change into a gas and steam starts to rise upwards from the spout. This is an example of a phase transition, a change of state of water. It is utterly discontinuous and strongly contrasts with the smooth continuous changes that have occurred previously.[7]

This is the sort of breathless discussion which characterises advocates of emergence. There is just one problem. Discontinuous phase transitions cannot exist in finite systems. Roughly, the proof goes like this. The energy of a molecule of water is a continuous function of temperature. Since the total energy of a body of water (its Hamiltonian) is simply a sum over the energy of all its component molecules, it too must be continuous. This applies to all other properties and so there can be no discontinuities.

All physical systems are finite. But phase transitions only exist in infinite systems. If you've just boiled a kettle or are sipping an iced drink, you might find this puzzling. The solution is, of course, that in finite systems phase transitions only appear discontinuous. This is same point I raised when I discussed boiling water in Chap. 8. If you have sufficiently sensitive instruments, you'll see a rich range of behaviour as water gradually changes to steam. Discontinuous phase transitions are a property of models in the thermodynamic limit, not of physical systems. Infinitely more may be different, but more is the same.

The idea that phase transitions are discontinuous is deeply rooted. Philosopher Chuang Liu speculates that this is self-reinforcing and wonders:

> ...why could we not say that since we now know that there are fluctuations in finite macrosystems, phase transitions in them should no longer be seen as singularities? It is quite possible that if such a conceptual shift takes place among theoreticians, the isotherms of systems with multiple phases coming back from the laboratories will no longer have singularities in them.[8]

Now let's turn to universality. Like phase transitions, strict universal behaviour only exists in the thermodynamic limit. Systems which share critical exponents in the limit will display different behaviour at finite scales. Physical systems may display approximate universality, remember the empirical work described in Sect. 8.7. But strict universality is a property of models.

Some see this as a characteristic of emergent phenomena:

---

[7] Blundell (2019), p. 238.
[8] Liu (1999).

...since there is no way to rigorously explain (i.e., derive mathematically) how a phase transition occurs in a finite, real system, then these phenomena do appear to be strongly emergent.[9]

or

An emergent behavior of a physical system is a qualitative property that can only occur in the limit that the number of microscopic constituents tends to infinity.[10]

Of course, you can define emergence in any way you want.[11] But defining it in terms of behaviour that is impossible in real physical systems seems peculiar, to say the least. It leads the authors of the second definition into a further quandary: "If, however, life and consciousness are sharply defined only in the thermodynamic limit, then we are only approximately alive and operationally conscious."[12]

The thermodynamic limit is used because it is an excellent approximation. To give an idea of how good the approximation is, let's take the example of temperature. In the thermodynamic limit temperature is constant without fluctuations. A one litre box of gas contains around 10^{22} molecules. At room temperature, random fluctuations in temperature are of the order of 10^{-10}K.[13] This is as close to zero as makes no differences for any practical purposes. Later in the chapter, I will give a similar estimate of how universal behaviour depends on the scale of the system.

It is also used because of the difficulty of the alternative. I mentioned Debye screening in Sect. 9.3. The derivation in the thermodynamic limit is a few lines of algebra. The recent derivation as a many-body system is a tour de force of theoretical physics.

But nature knows nothing of the thermodynamic limit and is always somehow solving many-body problems. Again, it seems proponents of emergence blur the distinction between reality and our models of reality. Here is philosopher Robert Batterman:

Such qualitative changes of state... cannot be reductively explained by the more fundamental theories of statistical mechanics. They are indeed emergent phenomena. The reason for this (rather dramatic) negative claim has to do with

---

[9] Bangu (2014), p. 162.

[10] Kivelson and Kivelson (2016).

[11] A similar definition is in Butterfield (2011).

[12] Kivelson and Kivelson (2016).

[13] The calculation is in Sect. 7.3.

the fact that such changes require certain infinite idealizations. From the point of view of the underlying fundamental theory whose proper focus is on the interactions of a finite number of molecular components of the macrosystems, these qualitative changes are genuinely novel. The upshot is that the statistical mechanics of finite systems is explanatorily insufficient.[14]

This reminds me how simulation emergence is defined in terms of the way in which a system is modelled and is subject to the same criticisms. It's not nature but our understanding that requires infinite idealisations. Currently, statistical mechanics may be insufficient, but this is a consequence of our cognitive and computational limitations. Just as with Debye screening, as soon as some smart people work out how to solve the relevant many-body problem, we will understand the system better.

The thermodynamic limit is not only a good approximation, but often there is no alternative to using it. While these two properties explain why it is so common, neither recommend it as the basis of a definition of emergence which should capture something about the world, not about models.

10.2 Effective Theories

When we write down theories, we focus on the things that matter at the scales we are interested in and ignore the rest. This happens throughout science. Newtonian mechanics isn't taught as the low velocity limit of Special Relativity, but as a topic in its own right. For everyday scales, it works just fine. Such theories are known as effective theories.

This applies to fundamental physics too. Since it consists of field theories, we talk of Effective Field Theories (EFTs). These only apply to a specific scale, ignoring interactions at larger and smaller scales. The scale could be of distance, time, or energy. We've already seen the example of Debye screening in Sect. 9.3 involving modelling one electron in a plasma as being subject to the effective field of all the other electrons. Another example is from nuclear physics. As long as we stick to relatively low energies, we can ignore that nucleons are made of quarks and gluons. At the other end of the scale, if we want to model the earth-moon system we can ignore gravitational interactions with the rest of the universe. The larger the separation between the scale of interest and the scales that are ignored, the more accurate the EFT will be.

---

[14] Batterman (2011).

EFTs are useful approximations and are found throughout physics, from condensed matter physics to cosmology to high energy physics. Indeed, the standard model of high energy physics is almost certainly itself an EFT, a low energy approximation to some as yet unknown higher energy theory. Just as in the discussion of levels in Sect. 3.1, it is impossible to know whether there is some fundamental theory. In his excellent review of EFTs, Howard Giorgi wrote:

> It is possible that the rules change dramatically, as in string theory.

> It may even be possible that there is no end, simply more and more scales as one goes to higher and higher energy.

> Who knows?

> Who cares?[15]

It's easy to see why EFTs are so much discussed in the emergence literature. They seem to give a formal structure to the idea of levels (one level = one EFT) and so pick out emergent phenomena. They automatically satisfy the definition of DOF emergence since they ignore DOFs at scales other than the one of interest. When we build an EFT, how do we know which scales to ignore? Bishop and co-authors claim that this is given by the context:

> Nothing in the 'fundamental theory' indicates how to make these choices. Even the most fundamental … theory must come to expression in a concrete context—a context not given by such a theory.[16]

Ellis goes further and asserts that EFTs are ontologically real in Wilson's sense of having causal powers:

> A key result is that all emergent levels are equally causally effective in the sense that the dynamics of every level L we can deal with in an empirical way is described by an Effective Theory ET_L at that level that is as valid as the effective theories at all other levels. There is no privileged level of causality.[17]

My response will come in several parts. The first is to note that EFTs don't somehow float free of lower levels. Instead, calculations based on lower levels

---

[15] Georgi (1993).
[16] Bishop et al. (2022), p. 209.
[17] Ellis (2020).

are often a fundamental part of their construction. The second is that once you look at the range of EFTs that exist, the assertion that they represent the structure of the world seems bizarre.

As an example, lets take EFTs arising from quantum chromodynamics, As we saw in Sect. 2.1, the nature of the strong interaction means the calculations become fiendishly difficult. This has spawned a whole host of effective theories, good enough approximations for specific problems. Figure 10.1 shows some of them. Who on earth could possibly think this represents the structure of reality?

Much of physics is about creating EFTs suitable for the particular property that is being studied. The result is a proliferation of EFTs; I could draw similar diagrams for cosmology or condensed matter physics.[19] What's more, as time passes, some of them will fall into disuse, some will be replaced by more accurate methods, others will remain part of the canon. This process has everything to do with the usefulness of a particular EFT to the current interests of scientists. It has little to do with the structure of reality.[20]

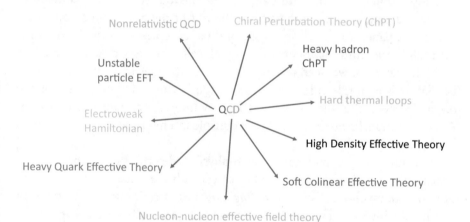

Fig. 10.1 The structure of reality?[18]

---

[18] Adapted from a diagram by Iain Stewart.

[19] For a review of the use of EFTs in cosmology, see Cabass et al. (2022) and in condensed matter physics, Brauner et al. (2022).

[20] Sabine Hossenfelder claims that EFTs may be an example of strong emergence: "There isn't a priori any reason why it must be possible to continue the constants of the theory at high resolution to any lower resolution. If you run into a point where the coupling can't be continued, you will need new initial values that have to be determined by measurement. Hence, strong emergence is viable", Hossenfelder (2019). It seems to me that this is a point about models, not about systems.

10.3   The Renormalization Group Transformation

The Renormalization Group Transformation (RGT) is an intimidating mouthful of a term. Yet it is one of the most important recent developments in physics. The central idea is relatively straightforward. It consists of a coarse graining transformation followed by a rescaling, or renormalization, of the system's parameters. We've already seen this sort of transformation as an example of multiple realization, shown in Fig. 8.1. To refresh your memory, that transformation mapped 9 cells in the original system onto one cell in the new. The leftmost two panels of Fig. 10.2 reproduces this, with the arrows showing the mapping.

Now imagine that the first panel shows part of a much larger grid. When we apply the transformation, the 6 × 6 grid maps on onto a 2 × 2 grid at the centre of the second panel, and the other cells in the second panel come from the parts of the original grid not shown. Then we can repeat the transformation to get the third panel, where the four central cells come from the second panel and the others from the rest of the grid. And so on to get the fourth panel. Each cell in the second panel represents nine in the first; each cell in the third represents 81 and each cell in the fourth represents 729.

With each step, we need to rescale whatever the metric of the system may be. Say it is a simple ruler, as shown at the bottom of the figure. For the original system, the ruler measures length, with each cell taken to be a unit. After each transformation, the scale is tripled. This process of rescaling is known as renormalization.

Now imagine we had some microphysical description of the original system in terms of the way one cell interacts with the others. Such a description would involve a list of *coupling constants* describing all the ways in which cells interact. This list will include interactions between neighbours,

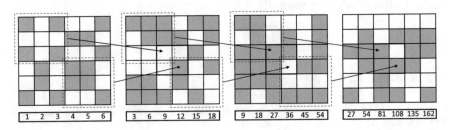

Fig. 10.2 Coarse graining

next-nearest neighbours etc. If this description is also valid after the transformation, perhaps with new values for the coupling constants, the system is *renormalizable*.[21]

This combination of a transformation which allows a subsequent renormalization gives the T and the R of RGT. It seems the G is there just to confuse non-mathematicians. Technically, the transformation is a semi-group which gives the G. However this property doesn't matter for our purposes.

That's the basic idea. The technique was first introduced in high energy physics during the 1950s. Kenneth Wilson was awarded the Nobel Prize in Physics in 1982 for his work on developing the theory and applying it to phase transitions. The RGT is so powerful because it can be applied to a whole range of many-body systems so has found applications across physics, from QCD all the way up to cosmology. For the purposes of this chapter, what matters is the way in which it gives an explanation for universality and the way in which this explanation is taken as support for emergence. To discuss this, we'll need to look at the RGT in a bit more detail. I'm going to keep the presentation free of maths. For a somewhat less fluffy treatment, see Appendix A.4. For a rigorous statement, I refer you to the Further Reading.

Let's start with an infinite 2-dimensional grid, the cells of which are either on or off. Then for the transformation let's use the coarse graining we've already seen in Fig. 10.2: each 3×3 block of cells is mapped onto a single cell. Apply the transformation repeatedly, zooming out to ever larger scales. Since we assumed the original grid was infinite, we can do this an infinite number of times. Then we can ask the question: after all these transformations, does the pattern of the cells stay the same? There are only 3 possible patterns which will remain unchanged:[22]

1. An *ordered* pattern with all the cells on or all the cells off.
2. A *disordered* pattern with the cells randomly on or off.
3. A *scale invariant* pattern.

In the first case, the grid is either all grey or all white so will not change with scale. In the second, the pattern remains random at any scale simply because the sum of random variables is also a random variable. The third case is more interesting. Scale invariance means that the pattern of fluctuations looks the same however many transformations we apply. For a pattern to

---

[21] There is a close connection between the RGT and the effective field theories of the previous section. Each application of the RGT can be thought of as producing an EFT at a larger scale.

[22] A link to a crystal clear video showing this process of zooming out, due to Doug Ashton, can be found at www.TheMaterialWorld.net.

be scale invariant, there must be fluctuations of all possible sizes, and their distribution must follow a power law.[23]

To relate these patterns to real phenomena, let's return to the mathematical description of the system. We saw in Sect. 6.6 that a many-body system is described by a Hamiltonian, a sum of the energies of all the elements. Given the fixed grid, the energy of a particular configuration will be determined by the nature of the interaction between the elements. If the cells represent magnets pointing up or down, the magnetic interaction tends to align neighbours. On the other hand, thermal noise will tend to randomise them. So the Hamiltonian of the system will depend on both the strength of the interaction and temperature.

We can use this to think about the three patterns. The first case corresponds either to a low temperature or a strong coupling constant: a solid phase. The second case to a high temperature or a weak coupling constant: a gaseous phase. The third case corresponds to a phase transition. Scale invariant patterns are not just a theoretical artefact but are observed in a range of systems. As a saucepan of water reaches boiling point you can see bubbles with a wide range of sizes. Such fluctuations result in the phenomenon of critical opalescence, first observed in the nineteenth century.

It's worth pausing here to appreciate how extraordinary this is. We started with a grid and have done nothing but think about its properties as we progressively look at it at larger and larger scales. Yet this has led to behaviour which seems characteristic of a whole range of physical systems. I can't think of a better example of the sheer wonder that simple models inspire.

More formally, these three cases are the fixed points of the transformation. The maths shows that most of the microscopic properties of the system are irrelevant for the properties of the fixed points. Specifically, the precise nature of the interactions doesn't matter: whether nearest neighbour, next-nearest neighbour or something more complex, all will have the same fixed points. This gives a definition of a universality class:

> ...members of a universality class have only three things in common: the symmetry group of the Hamiltonian, the dimensionality and whether or not the forces are short-ranged.[24]

This is the first part of the RGT's explanation of universality: the fixed points are the same for a wide range of systems. Then we can ask how

---

[23] See Appendix A.4 for the proof. A link to a video showing scale invariance, due to Doug Ashton, can be found at www.TheMaterialWorld.net.

[24] Goldenfeld (1992), p. 80.

the system behaves in the vicinity of the fixed points. This can be studied by linearising around the fixed point (linearisation was briefly discussed in Sect. 7.3). It turns out that the dynamics depend only on the symmetry properties of the Hamiltonian and not on the details of the interactions. This can be used to derive the critical exponents characteristic of universal behaviour. This is the second part of the explanation of universality.

It's not hard to see why proponents of emergence get excited about this. Each step of the transformation adds another level of MR emergence until in the limit all that is left is universality. In terms of DOF emergence, reducing degrees of freedom is exactly what the transformation does. In terms of contextual emergence, the RGT shows that you need to look at the system at the largest scales to understand behaviour at the smallest. Here are a couple of quotes:

> The renormalization group [is] ... inherently multi-scale. They are not bottom-up derivational explanations.[25]

> One can think of the application of the RG as throwing away superfluous information from the smallest scales, while preserving the behavior that matters independent of the small-scale details. Indeed, RG methods allow the construction of a hierarchy of effective theories at larger length and longer time scales without knowing the details of the physics at the smallest length and time scale (if there is a "smallest"). This fits with the theme of this chapter—namely, that there is no "fundamental" scale or "fundamental" entities determining everything else at larger length and longer time scales.[26]

I want to convince you that all this is mistaken. I will do so using three arguments. The first is that the RGT is a model. The transformation itself is an abstract mathematical operation that bears no relation to any physical process. Nothing like coarse graining happens in a physical system. Throwing away detail is a correct description of this mathematical process but tells us nothing about what happens in the world. Physical systems do not, and indeed cannot, throw away detail.[27]

---

[25] Batterman (2018).

[26] Bishop et al. (2022), p. 142.

[27] This is not strictly true. Coarse graining requires information processing. Given what we know about chemical computation (Sect. 9.6), it wouldn't be a big step to design a reaction network with a threshold, which would be the most primitive form of coarse graining. Living systems include more complex computational networks, but the principle is the same. Throwing away detail is exactly what is going on when humans use concepts to describe the world. Flack (2017) is an interesting discussion of coarse-graining in biological systems though I disagree with the paper's conclusions about causation.

You might retort that the process is fundamental to the RGT's explanation of universality. This leads to my second point. The explanation tells us that systems whose Hamiltonians have a particular form will belong to the same universality class. But a Hamiltonian is a microscopic description. To write it down we need to carefully analyse the way that one element interacts with its neighbours. For a fluid, this will involve the various types of intermolecular force. For magnetism the exact nature of magnetic interactions between atoms. If that's not microphysics, I don't know what is.

What's more, talk of throwing away detail misses the point. The transformation is exact and the system after the transformation fully determined by the original system. The only details that are thrown away are the arbitrary configuration, whether this cell is on or that cell is off. This is a bit like saying the height of the tree from which Newton's apple fell was thrown away. It's true, but irrelevant to the universal law of gravitation.

My third point relates to the thermodynamic limit. The RGT involves two assumptions of infinity: that the system is infinite and that the transformation is applied an infinite number of times. It's not hard to see why this is necessary. For a finite system, progressively zooming out will at some point hit the boundaries of the system so there is no possibility of a fixed point.

As discussed in the first section, when we turn to real systems that are large but finite, the results of RGT will only be approximate. Goldenfeld gives a neat way of estimating the magnitude of the approximation. The scale invariance of fluctuations at the critical point means that there are fluctuations at all scales, but in a real system, fluctuations cannot be bigger than the size of the system. For a system of 1cm, he shows that this constraint will only matter within around 10^{-9}K of the critical point.[28] Further, the RGT result about universality is strictly local. Even in the infinite limit, universality only applies exactly at the fixed points. Away from the fixed points, all the microscopic details will affect the systems behaviour. This will be exacerbated in finite systems.[29]

Universality is a property of models. In the world, universality is only approximate. This is rarely stated. An exception is Wikipedia:

> In statistical mechanics, a universality class is a collection of mathematical models which share a single scale-invariant limit under the process of renormalization group flow. While the models within a class may differ dramatically

---

[28] See Appendix A.4.

[29] This can be investigated within the framework of RGT using a process called finite-size scaling. See Goldenfeld (1992), p. 279.

at finite scales, their behavior will become increasingly similar as the limit scale is approached.[30]

There are many fascinating aspects to RGT but it seems to me that none of them have much to do with emergence. On the contrary, RGT provides a reductive justification of why simplified models can be applied to complex systems, a reductive explanation of universal behaviour and a reductive explanation of the approximations involved in both.

10.4   Example: The Ising Model

The Ising model was originally posited in the 1920s as a mathematical model of ferromagnetism. Since then, it has come to be the canonical model of phase transitions. The set up is straightforward. Magnets, or spins, which can point either up or down are arranged on a lattice. The grids I've used both in the discussion of multiple realization and in the previous section can be taken to represent the model, with cells being on or off corresponding to spins being up or down.

We are interested in the overall magnetisation of the system. There are two offsetting effects. Neighbouring magnets tend to align since their energy will be lower if they do so. But thermal noise tends to randomise their orientation. We'd expect the first effect to dominate for low temperatures resulting in a non-zero magnetisation and the second to dominate at high temperatures, resulting in a magnetisation of zero.[31]

A closed-form expression for the magnetisation was first given in 1948 by Lars Onsager. A graph of Onsager's expression is shown in Fig. 10.3. Temperature is on the x-axis and the average magnetisation per site on the y, the units are unimportant.[32] Starting at low temperature, all the spins are either aligned upwards, giving an overall magnetisation of + 1, or all aligned downwards, giving a magnetisation of − 1. As the temperature increases, thermal noise starts flipping some spins out of alignment but the effect is too small to see on the graph. At a temperature of between 1 and 2, you can start to see the average magnetisation start to decrease and then fall sharply to zero. This is the phase transition, from a state with non-zero magnetisation to a state with zero magnetisation. It happens at a critical temperature

---

[30] https://en.wikipedia.org/wiki/universality_class.

[31] More details are in Appendix A.5.

[32] The Hamiltonian depends on the strength of the interaction between the spins and the Boltzmann constant. Following convention, to obtain the units shown in the figure I normalize both to unity.

which Onsager showed to be T = 2.27, in the same units as used for the figure.[33] Unsurprisingly, the derivation assumes the thermodynamic limit so can involve discontinuities. You can see the discontinuity in the graph when the magnetisation changes from having a near-vertical slope to a zero slope at the critical temperature.

Now imagine starting on the right of the graph and gradually reducing the temperature. When the critical temperature is reached, whether the final magnetisation ends up as + 1 or − 1 will be effectively random. This is an example of the symmetry breaking discussed in Sect. 2.5.

Let's now apply the RGT.[34] The maths is quite complex, but the steps are as follows:

1. Choose a coarse graining transformation.
2. This will imply a set of renormalized coupling constants.
3. Impose the condition that the Hamiltonian for the transformed system must have the same form as the original. This gives an expression relating the new coupling constants to the original ones.

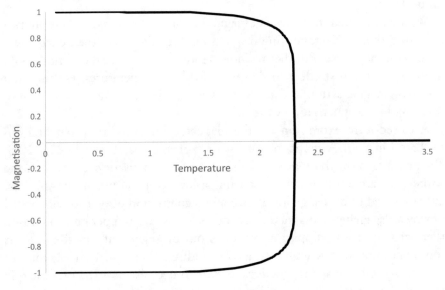

Fig. 10.3 2D Ising model magnetisation

[33] Onsager's expression for the critical temperature is astonishing: $2/\ln\left(1 + \sqrt{2}\right)$.

[34] This is taken from Goldenfeld (1992), Sect. 9.6.

4. This expression will be difficult to solve. So start with a first-order approximation. This will give an approximate expression that shows how the coupling constants are transformed.
5. Solve this for the fixed points. The values (in terms of temperature) are $T = 0$ and $T = \infty$ corresponding to the low and high temperature fixed point, and a third which corresponds to the critical temperature $T_C = 2.9$. This is not far from the exact value of 2.27.
6. This is one part of universality. The other is the dynamics near the fixed point which are determined by an eigenvalue. Calculating this from the approximation gives 1.62 compared to the exact value of 1.73.

Pause for a moment to reflect on this. The Ising model is about as simple as you can get, abstracting away from almost all the detail of real physical systems. Yet the last section showed it can give qualitative properties of the properties of phases. This is already remarkable. But to get quantitative predictions for the critical exponent which characterises phase transitions is staggering.

After working through the complex derivation, coding up the numerical solution comes as a relief. The system is easy to simulate,[35] involving just a few steps

1. Define a 2D matrix, each element of which represents a spin
2. Pick a temperature

   a. For each element, calculate its energy as a sum over the relative alignments of its 4 neighbours
   b. For the chosen temperature, calculate the probability of thermal noise randomly flipping the spin
   c. Take a random draw between 0 and 1. If it is higher than the probability, flip the spin
   d. Repeat this for all spins
   e. Go back to step (a) and repeat for the chosen number of periods

3. Go back to step (2) and repeat for a different temperature.

I mentioned in Sect. 5.1 that most simulations involve fudges. This one is no exception. The procedure I just described works but is quite slow. To speed it up, it is better to flip a small proportion of spins, the value I use is

---

[35] The code is available at www.TheMaterialWorld.net. It runs using the excellent open-source program Octave.

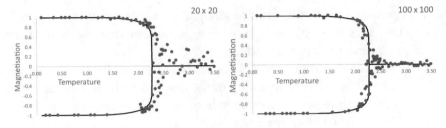

Fig. 10.4 Magnetisation in the Ising model

10%, at each iteration. This doesn't change the results but means the code runs an order of magnitude faster.[36]

The results of the simulation can be studied in many different ways.[37] Figure 10.4 shows how the absolute value of magnetisation varies with temperature for 20 × 20 and 100 × 100 lattices. The line shows the theoretical result (as in Fig. 10.3) and the dots are the results of simulation. Two things stand out. First, the phase transition is not sharp but smeared out over a range of temperature. Second, for the larger lattice, there is less smearing out and the points are generally closer to the theoretical value.

To look at this more formally, you can fit a smooth curve to the points from the simulation. This was done in 1976 by David Landau in what has become a classic paper in the simulation literature.[38] Figure 10.5 is taken from this paper. Again, this plots magnetisation against temperature, with the theoretical value shown by the dashed line. The other lines show magnetisation for lattices of different sizes. The highest line, marked by empty circles, is the smallest lattice; the lowest line, with filled circles, is the largest. You can see how the transition gets steeper and steeper as the lattice size gets larger. Turn this round: the larger the lattice size, the better an approximation is the thermodynamic limit.

Here's what Leo Kadanoff, whose work on block spins underlies the RGT approach, has to say about such a figure:

> As the number of lattice sites gets larger the variation in the magnetization will get steeper, until at a very large number of sites the transition from positive values of ⟨σ⟩ to negative ones will become so steep that the casual observer might say that it has occurred suddenly. The astute observer will look

---

[36] This is because programs such as Octave or Matlab are optimized for matrix operations. Flipping cells one at a time requires going through a matrix element by element so is inefficient.

[37] The data in the figures comes from runs with 20,000 periods flipping 10% of the spins each period. I have excluded *metastable* states which have close to zero magnetization even at low temperatures, as these are not relevant to my argument. If you are interested, download the code.

[38] Landau (1976).

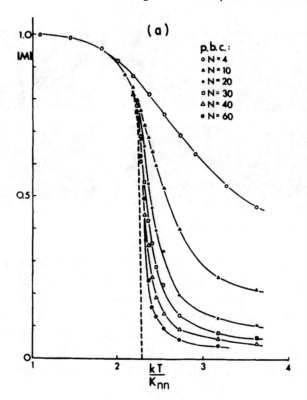

Fig. 10.5 How magnetisation changes with lattice size[39]

more closely, see that there is a very steep rise, and perhaps conclude that the discontinuous jump only occurs in the infinite system.[40]

This captures exactly the sense in which the thermodynamic limit is a useful approximation and that phase transitions in real systems appear continuous to a casual observer, or at least one without extraordinarily high-resolution instruments.

To further study these results, Fig. 10.6 shows a selection of the lattices from the 100 × 100 case.[41] Temperature increases from left to right, first along the top row then along the bottom row. The lattice at the top left is at a low temperature. This solid phase has all the spins aligned; hence the uniform

---

[39] Reprinted Fig. 10.4 with permission from Landau (1976). © 1976 by the American Physical Society.
[40] Kadanoff (2009).
[41] A video is available at www.TheMaterialWorld.net.

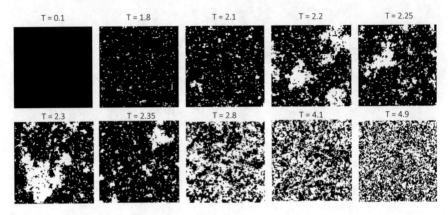

Fig. 10.6 A phase transition, step by step

colour. As we move to the left and the temperature increases, more and more spins and flipped randomly and you can see white patches appearing.

The rightmost two panels on the top row and the leftmost two on the bottom row show lattices near the critical temperature. What is striking is that you can see how the phase transition happens. Fluctuations appear at all scales, with some filling large parts of the grid. As the temperature increases above its critical value, the rightmost three panels on the bottom row show these fluctuations collapsing into randomness.

What does all this have to do with the universality in real systems described in Sect. 8.4? For this, we need a 3D Ising model which falls into the same universality class. No closed-form results exist, but it is straightforward to analyse the equations numerically or extend the simulation to 3D. If this is done, it turns out that the critical exponent[42] is 0.31, which is close to the value of 1/3 used in Fig. 8.2 which shows the empirical results on universality. It is remarkable that such different systems show such similar properties. But similarity is all it is; strict universality is only found in the thermodynamic limit.

10.5 Models and Emergence

If you are looking for examples of the different types of emergence, the Ising model is a good place to start. The 2D version has a closed-form solution; the 3D version doesn't so one is simulation emergent and the other not. You can find MR emergence wherever you want, zooming out to some level and

---

[42] The value is taken from Chandler (1987), p. 123.

studying its properties. If you're looking for contextual emergence, it is indeed impossible to explain the behaviour of a few cells without referring to the behaviour of the whole system. The model displays a phase transition and is a canonical example of universality.

Yet in the previous section I did not use the term emergence. Instead I use standard tools, closed-form results, approximations or simulations, albeit presented in a non-technical way. These allow a thorough understanding of the Ising mode and shows how so-called emergent properties are the result of approximations and allows them to be understood in purely bottom-up terms. Even the randomness off symmetry breaking as the lattice cools is only apparent. Given thermal noise (or more precisely the sequence of pseudo-random numbers in the code that represents thermal noise), if you are prepared to put in the effort you can always unravel the causal chain that leads to a configuration having a positive or negative magnetisation at low temperatures.

Universality is a fascinating concept and its explanation by RGT is a jewel of theoretical physics. But, in real systems, it is always an approximate one. The RGT tells us what properties systems need to share if there are to belong to the same universality class. If we want to understand this for a particular system, we need to do the hard work of reductive physics, analyse carefully the nature of intermolecular interactions and explain why they satisfy the conditions suggested by the RGT. This is only part of the problem. Then we need to study the exact physical processes that occur around the phase transition, in the style of the paper from which I took the image of boiling water in Fig. 8.3.

A first go at comparing phase transitions in the Ising model and water might look like this. As temperature rises, the amplitude and frequency of fluctuations increases. In the Ising model, this leads to random patches of magnetisation. The interior of these patches is relatively stable against fluctuations, since each spin is surrounded by others by the same orientation. This means that the patches tend to grow. Since these patches are first seen well away from the critical point, by the time it is reached there are patches at all scales with larger patches being older. Above the critical point, the temperature becomes high enough to disrupt even the inside of patches so the pattern quickly collapses. For water, temperature fluctuations create tiny bubbles of steam. As the temperature rises, the steam's pressure increases and so the bubbles become bigger. From then on, the story is the same as for the Ising model except that at the critical point the steam and water separate. Far from universality excluding reduction, as proponents of emergence claim, only reduction can explain universality.

10.6 Minimal Models, Sloppy Models and Multiphysics

So far, the models I've described are *minimal models*. This term was introduced by Goldenfeld in 1993. Here is his more recent description:

> Gratuitously realistic details of the starting model will not only complicate the technical task of extracting the asymptotic structure, but will not impact the **leading order behavior**. Thus, a minimal model represents a universality class of models; whether or not one uses the minimal model or a decoration of it, the **asymptotic outcome** of a nonperturbative calculation will not be different.[43]

The RGT explains why minimal models work. Systems which share some broad properties belong to a universality class and show similar behaviour around fixed points. A minimal model can be thought of the simplest system in a universality class. The quote makes it clear that this is only valid in the thermodynamic limit ("asymptotic outcome") and that it is an approximate result ("leading order behaviour"). Such models are key to our imaginative understanding, they provide what Anderson describes as the "enormous compression of the brute-force calculational algorithm, down to a set of ideas which the human mind can grasp as a whole".[44]

James Sethna's idea of a *sloppy* model is closely related. A sloppy model is dependent on a large number of parameters but its predictions are insensitive to most of them. Proponents of the idea argue that it explains the success of minimal models, effective field theories and simplified modelling in general. All these share the characteristic of ignoring the sloppy parameters and focussing on those that matter. You might recognise this as a restatement, in modelling terms, of multiple realization. And it is vulnerable to the same criticism. The predictions of models might be insensitive to many parameters, but this is because they are approximations.

Minimal models are at one extreme of the modelling spectrum. At the other are multiphysics (or multiscale) models. These involve modelling systems at different levels, using physics appropriate to each, then linking the different levels to allow them to interact.

Many models of molecular systems make use of QM / MM models, which treat some details of large molecules using exact but computational expensive

---

43 Goldenfeld (2024), my emphasis.
44 Anderson (2011), p. 136.

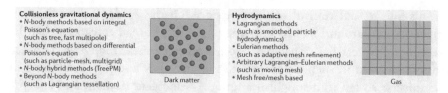

Fig. 10.7 Multi-physics and the early universe[45]

quantum mechanics (QM) and the rest using approximate but computationally easier classical molecular mechanics (MM). For example, when modelling an enzyme you might use QM to model the active site, where catalysis happens, and MM for the rest of the molecular structure.

The simulations I briefly mentioned in Sect. 5.1 take a similar approach, blending different sorts of model. One of these was a simulation of the early universe. Broadly, it models weakly-interacting dark matter using many-body techniques while ordinary matter is modelled using fluid dynamics and then two are coupled together. Figure 10.7, taken from the paper, gives an idea of the range of techniques used.

Further examples can be found from materials science to geology and biology to engineering.[46] Bishop et al. claim that "multiscale modelling is an application of the contextual emergence framework and is best understood in that light".[47] To me, it seems more an example of making whatever assumptions are necessary to get something fit for purpose. These assumptions will depend on the budget, timescale and computing resources available. It's hard to imagine being further from the nature of the world.

10.7 Discussion

The chapter on MR emergence ended with this quote: "We cannot simply read the causal structure of the world just by looking at what nouns scientists use".[48] The conclusion of this chapter can be stated in similar terms: we cannot read off the structure of reality just by looking at the models scientists use.

---

[45] Source: Vogelsberger et al. (2020). https://www.nature.com/articles/s42254-019-0127-2. Reproduced with permission from Springer Nature.

[46] Respectively, Fish et al. (2021); Regenauer-Lieb et al. (2013); McCulloch (2016); https://www.comsol.com/comsol-multiphysics.

[47] Bishop et al. (2022), p. 84.

[48] Klein (2008).

This finishes our tour of weak emergence. Of the 75 definitions listed in Appendix A.1, around half fit into one of the three broad categories that the preceding chapters have analysed. None comes close to resolving the basic tension involved in such a position. I've argued they apply to everything in the universe so fail to distinguish a uniquely emergent domain and collapse into physicalism.

Without weak emergence, there is no possibility of non-reductive physicalism. You cannot have your physical cake and eat it with the pleasure of knowing it is beyond physics. All this is another way of saying that Kim's causal exclusion argument (Sect. 3.4) turns out to be correct. Remember, the argument doesn't exclude emergence, but says that if it exists it must be strong emergence, violating physical causal closure. That is the subject of the next chapter.

10.8   Further Reading

Introductions to thermodynamics can be found in many textbooks, my favourites are Waldram (1985) and Steane (2017). For an interesting discussion of the thermodynamic limit by a physicist see Styer (2004) and by a philosopher, Callender (2001). Behaviour at finite scales is covered in Franklin (2019) and Sethna (2022). Wilde (2021) studies phase transitions in nanoscale systems.

A good overview of EFTs can be found in Bain (2013) and a book length treatment in Burgess (2021). An excellent discussion of the relation of EFTs to emergence is in Luu and Meißner (2022) and it is worth reading Ellis's response in Ellis (2022).

The standard reference on the Renormalization Group is Goldenfeld (1992). A short overview can be found in Hüttemann et al. (2014) along with an interesting discussion of its implications for emergence. For applications of the RGT across the range of physics, see Huang (2013).

For minimal models, see Goldenfeld (2024), for sloppy models Gutenkunst et al. (2007) and for an application of QM / MM methods to modelling DNA, Bacolla et al. (2015).

More suggestions for reading can be found at www.TheMaterialWorld.net.

References

Anderson PW (2011) More and different: notes from a thoughtful curmudgeon. World Scientific, New Jersey

Bacolla A, Zhu X, Chen H, et al (2015) Local DNA dynamics shape mutational patterns of mononucleotide repeats in human genomes. Nucleic Acids Research 43:5065–5080. https://doi.org/10.1093/nar/gkv364

Bain J (2013) Effective Field Theories. In: The Oxford handbook of philosophy of physics. Oxford University Press, Oxford

Bangu S (2014) Neither Weak, Nor Strong? Emergence and Functional Reduction. In: Why more is different. Springer, New York

Batterman RW (2011) Emergence, Singularities, and Symmetry Breaking. Found Phys 41:1031–1050. https://doi.org/10.1007/s10701-010-9493-4

Batterman RW (2018) Autonomy of Theories: An Explanatory Problem. Noûs 52:858–873. https://doi.org/10.1111/nous.12191

Bishop RC, Silberstein M, Pexton M (2022) Emergence in context: a treatise in twenty-first century natural philosophy. Oxford University Press, Oxford, United Kingdom

Blundell SJ (2019) Phase transitions, broken symmetry and the renormalization group. In: Gibb SC (ed) The Routledge handbook of emergence. Routledge, New York

Brauner T, Hartnoll SA, Kovtun P, et al (2022) Snowmass White Paper: Effective Field Theories for Condensed Matter Systems. https://doi.org/10.48550/ARXIV.2203.10110

Burgess CP (2021) Introduction to effective field theory: thinking effectively about hierarchies of scale. Cambridge University Press, Cambridge.

Butterfield J (2011) Less is Different: Emergence and Reduction Reconciled. Found Phys 41:1065–1135. https://doi.org/10.1007/s10701-010-9516-1

Cabass G, Ivanov MM, Lewandowski M, et al (2022) Snowmass White Paper: Effective Field Theories in Cosmology. https://doi.org/10.48550/arXiv.2203.08232

Callender C (2001) Taking Thermodynamics Too Seriously. Studies in History and Philosophy of Science Part B: Studies in History and Philosophy of Modern Physics 32:539–553. https://doi.org/10.1016/S1355-2198(01)00025-9

Cariani P (1991) Emergence and artificial life. In: Langton CG, Taylor CE, Farmer JD, Rasmussen S (eds) Artificial Life II. Addison-Wesley, Reading, MA

Chandler D (1987) Introduction to modern statistical mechanics. Oxford University Press, New York Oxford

Ellis G (2020) Emergence in Solid State Physics and Biology. Found Phys 50:1098–1139. https://doi.org/10.1007/s10701-020-00367-z

Ellis G (2022) Response to Part II: The View from Physics. In: Top-Down Causation and Emergence. Springer Nature, S.l.

Fish J, Wagner GJ, Keten S (2021) Mesoscopic and multiscale modelling in materials. Nat Mater 20:774–786. https://doi.org/10.1038/s41563-020-00913-0

Flack JC (2017) Coarse-graining as a downward causation mechanism. Phil Trans R Soc A 375:20160338. https://doi.org/10.1098/rsta.2016.0338

Franklin A (2019) Universality Reduced. Philosophy of Science 86:1295–1306. https://doi.org/10.1086/705473

Georgi H (1993) Effective Field Theory. Annu Rev Nucl Part Sci 43:209–252. https://doi.org/10.1146/annurev.ns.43.120193.001233

Goldenfeld N (1992) Lectures on phase transitions and the renormalization group. Addison-Wesley, Reading (Mass.)

Goldenfeld N (2024) There's Plenty of Room in the Middle: The Unsung Revolution of the Renormalization Group. In: 50 Years of the Renormalization Group: Dedicated to the Memory of Michael E Fisher. World Scientific

Gutenkunst RN, Waterfall JJ, Casey FP, et al (2007) Universally Sloppy Parameter Sensitivities in Systems Biology Models. PLoS Comput Biol 3:e189. https://doi.org/10.1371/journal.pcbi.0030189

Hilbert D (1998) On the infinite. In: Benacerraf P, Putnam H (eds) Philosophy of mathematics: selected readings. Cambridge University Press, Cambridge

Hossenfelder S (2019) The Case for Strong Emergence. In: Aguirre A, Foster B, Merali Z (eds) What is Fundamental? Springer International Publishing, Cham, pp 85–94

Huang K (2013) A critical history of renormalization. Int J Mod Phys A 28:1330050. https://doi.org/10.1142/S0217751X13300500

Hüttemann A, Kühn R, Terzidis O (2014) Stability, Emergence and Part-Whole Reduction. In: Why more is different. Springer, New York

Kadanoff LP (2009) More is the Same; Phase Transitions and Mean Field Theories. J Stat Phys 137:777–797. https://doi.org/10.1007/s10955-009-9814-1

Kivelson S, Kivelson SA (2016) Defining emergence in physics. npj Quant Mater 1:16024. https://doi.org/10.1038/npjquantmats.2016.24

Klein C (2008) An ideal solution to disputes about multiply realized kinds. Philos Stud 140:161–177. https://doi.org/10.1007/s11098-007-9135-7

Landau DP (1976) Finite-size behavior of the Ising square lattice. Phys Rev B 13:2997–3011. https://doi.org/10.1103/PhysRevB.13.2997

Lanford OE (1975) Time evolution of large classical systems. In: Moser J (ed) Dynamical Systems, Theory and Applications. Springer, Berlin, pp 1–111

Liu C (1999) Explaining the Emergence of Cooperative Phenomena. Philos of Sci 66:S92–S106. https://doi.org/10.1086/392718

Luu T, Meißner U-G (2022) On the Topic of Emergence from an Effective Field Theory Perspective. In: Top-Down Causation and Emergence. Springer Nature, S.l.

McCulloch AD (2016) Systems Biophysics: Multiscale Biophysical Modeling of Organ Systems. Biophysical Journal 110:1023–1027. https://doi.org/10.1016/j.bpj.2016.02.007

Regenauer-Lieb K, Veveakis M, Poulet T, et al (2013) Multiscale coupling and multiphysics approaches in earth sciences: Theory. j coupled syst multiscale dyn 1:49–73. https://doi.org/10.1166/jcsmd.2013.1012

Sethna JP (2022) Power laws in physics. Nat Rev Phys 4:501–503. https://doi.org/10.1038/s42254-022-00491-x

Steane AM (2017) Thermodynamics: a complete undergraduate course. Oxford University Press, Oxford

Styer DF (2004) What good is the thermodynamic limit? American Journal of Physics 72:25–29. https://doi.org/10.1119/1.1621028

Vogelsberger M, Marinacci F, Torrey P, Puchwein E (2020) Cosmological simulations of galaxy formation. Nat Rev Phys 2:42–66. https://doi.org/10.1038/s42254-019-0127-2

Waldram JR (1985) The theory of thermodynamics. Cambridge University Press, Cambridge

Wilde G (2021) Structural Phase Transformations in Nanoscale Systems. Adv Eng Mater 23:2001387. https://doi.org/10.1002/adem.202001387

11

Strong Emergence

Summary Strong emergence means that some things are beyond current physics. This could happen in various ways. A higher level feature could have genuine causal power and violate physical causal closure. Or it could affect the lower level by exploiting unknown physics or the causal slack left by the indeterminism of quantum physics. Or particular configurations of matter could evoke something that is non-physical. There is no logical reason to exclude any such hypotheses, but this chapter argues that neither is there any convincing empirical or theoretical evidence which makes them worth entertaining.

The last chapters showed that weak emergence is so weak that it applies to all systems hence tells us nothing about the world. If you want to continue to claim the reality of emergent properties, you need to accept that they contradict current physics and so are strongly emergent.

Recall Wilson's definition: strongly emergent phenomena have causal powers over and above their physical base so microphysics must conform to the higher level. This could happen in various ways. One is as a result of new physics. Another is by exploiting causal slack at the fundamental level, which would require an explanation of how physical causal closure permits such slack. Another is by simply arguing that physics does what the higher level tells it to. Then there are concepts of strong emergence related to computability which argue not that physical causal closure is violated but rather that some things are outside physics.

L. Graham, *Physics Fixes All the Facts*, The Frontiers Collection, https://doi.org/10.1007/978-3-031-69288-8_11

This chapter is a bit of a rag-tag of ideas that don't have much in common apart from loosely satisfying the definition of strong emergence. This is the case even though I am going to neglect the large body of work which proceeds as follows:

(i) Identify a phenomenon
(ii) Describe at length its complexity or the extraordinary nature of its properties
(iii) Assert that it is obvious that it cannot be reduced to physics so it must be strongly emergent.

Claims for strong emergence in mental causation and consciousness fall into this category.[1] So does a range of more interesting work which investigates biological examples from cellular mechanisms, to evolutionary transitions to the role of the epigenome.[2] However, it seems to me that without a careful explanation of the mechanism by which physical causal closure is violated or sidestepped, these add up to little more than saying "Wow! There's no way that all that complexity can be just quantum physics!".

11.1 New Physics

Physics involves four fundamental interactions: electromagnetism, the strong and weak interactions, and gravity. The first three are described by quantum physics. Gravity is described by general relativity. Both theories have been tested to extraordinary levels of accuracy. A 2022 paper[3] showed that one implication of quantum physics matched experiment to 13 digits of accuracy. In the same year, another paper[4] reported a test of general relativity to 15 digits.

These accuracies can be seen as constraints on any proposals for new physics. Few physicists would doubt that new physics is needed: at high energy levels, quantum physics and general relativity are mutually inconsistent; the measured expansion of the universe is different from predictions and there are regular hints that something is missing from the standard model. Proposals to address these issues involve extremely high energies and either extremely large or extremely small scales. But almost all of the phenomena

---

[1] Examples are Deacon (2012) and Clayton (2006).
[2] Respectively, Mcleish (2017), Walker et al. (2012) and Davies (2012).
[3] Sailer et al. (2022).
[4] Touboul et al. (2022).

that are described as emergent exist at our low energy, mesoscale world. Arguing that some future theory of quantum gravity would have implications for such systems would be challenging, to say the least.[5]

This difficulty may explain the rarity of such proposals in the recent emergence literature. If we look further back, in the first half of the twentieth century the British Emergentists spoke of configurational forces that are fundamental but only experienced by particular arrangements of matter.[6] Atoms arranged randomly would not experience them. But arrange them into, say, a crystal, and a new force somehow switches on. There is no evidence for such forces.

To find candidates for strong emergence we need to look at the wilder boundaries of science. One example that comes to mind is Rupert Sheldrake's idea of morphic resonance which was widely discussed in the 1980s.[7] This was the hypothesis that patterns are somehow reinforcing and Sheldrake was thinking particularly in terms of the growth and evolution of organisms. He proposed several empirical tests. One which has stuck with me is that it should be easier to solve a newspaper crossword puzzle the day after its publication since, as people discover the solution, it becomes integrated into the morphic field. However none of the predictions were confirmed and the hypothesis can be dismissed. But it fits the model for strong emergence: high level behaviour (biological forms, crossword puzzle solutions) being determined by the morphic field and this somehow influencing the underlying physics.

A more recent example is Melvyn Vopson's mass-energy-information equivalence principle.[8] This asserts that information has mass and estimates the mass of one bit of information at room temperature to be around 10^{-38}kg. Apart from its interesting implications,[9] the hypothesis also seems to be an example of strong emergence. It implies that a high level feature (the arrangement of bits into states of higher or lower information content) affects physics (the mass of the system).

Such proposals for strong emergence should make the heart of a hardcore physicalist sing. Imagine if Vopson's proposal finds support from some unconnected part of physics. Then it is not hard to imagine a thorough effort to test it empirically. If it were verified, a whole army of theoreticians would

---

[5] Roger Penrose's work on the role of gravity in the quantum collapse process and the role of quantum effects in the brain is an exception, see Penrose (1999).

[6] For configurational forces and British Emergentism, see McLaughlin (2008).

[7] Sheldrake (1995).

[8] Vopson (2019).

[9] Vopson (2022).

start trying to work out how it fitted into fundamental physics. The scope of physics would be expanded. Physical causal closure would not be violated but extended. Strong emergence, in this sense, would be nothing but future physics.

11.2 Entanglement

Quantum physics and strong emergence could be linked in two ways. This section investigates quantum phenomena which fit the definition of strong emergence. The next askes whether quantum physics can explain how higher levels can have causal effects.

Before starting, let me give a word of warning. While the formalism of quantum physics is unambiguous (hence the accuracy of its predictions), its interpretation is an open question. Wikipedia lists 13 "influential" interpretations and adding in "minority" ones brings the count to around 20.

Most physicists don't worry too much about interpretations and just get on with using the equations of quantum physics to explain the world. Here is Stephen Hawking:

> I don't demand that a theory correspond to reality because I don't know what it is. Reality is not a quality you can test with litmus paper. All I'm concerned with is that the theory should predict the results of measurements. Quantum theory does this very successfully.[10]

Or more succinctly, and possibly apocryphally: "When I hear of Schrödinger's cat, I reach for my gun". The links between strong emergence and quantum physics in this section depend on particular interpretations and those in the next section involve new interpretations. In the absence of a way of deciding between interpretations, and in my opinion no one has a clue of how to do so, all this discussion remains hypothetical.

So let's start. A fundamental property of quantum physics is entanglement. Particles that interact become entangled and are described by a single wavefunction. If these particles become separated, they can influence each other apparently instantaneously, in any case faster than permitted by special relativity[11]. This is the implication of the famous Einstein Podolosky Rosen

---

[10] Hawking and Penrose (2000), p. 121.

[11] Relativistic causation is not violated since the faster than light interaction cannot be used to transmit information. The terminology is a bit of a mess here, since in quantum field theory "non-local" is used to describe processes that violate relativistic causation and "non-separable" to describe processes such as entanglement.

(EPR) experiment,[12] proposed in 1935 as a thought experiment to show quantum physics must be incomplete. The experiment was first carried out in 1982 by a team lead by Alain Aspect, and the predictions of quantum physics were fully verified.[13] Since then, experimental techniques have been progressively refined and have continued to confirm Aspect's result

If two photons are entangled on earth and sent off to opposite ends of the universe, they remain entangled forever, however great the distance between them. The entangled state determines the property of the individual particles. Information about just one part of an entangled system is not enough to know how the part will behave: for that we need information about the whole system. This sounds like strong emergence.[14] However, it does not pose a challenge for physicalism since it is physics. Entanglement shows that quantum physics is fundamentally non-local.[15] But the equations of quantum physics apply to the entangled system.

A paper[16] from 2018 describes a simple quantum system and claims it shows strong emergence in two senses. First, the system as a whole determines the properties of the individual components. Second, it is impossible in principle to derive the properties of the system from those of its components:

> ...the higher-order (i.e., N -particle) correlations are required to construct any of the lower-order correlations, but we cannot go the other way around; i.e., we cannot deduce the N -particle correlations from any of the lower-order correlations[17]

The paper's key assumption is that of weak measurement. Measurement always disturbs the quantum system. A proof is due to Paul Busch and is known as Busch's theorem.[18] A weak measurement is one that only weakly disturbs the system being measured. The price for this is that little information is extracted:

> The intuitive idea behind the weak value of an observable is that by letting the strength of the (standard) interaction between the object and probe become sufficiently weak, the disturbance caused by the first measurement on the

---

[12] Einstein et al. (1935).

[13] Aspect et al. (1982).

[14] Silberstein and McGeever (1999) make the case for entanglement being ontologically emergent.

[15] This is dependent on the interpretation, for example superdeterministic interpretations restore locality. For a review, see Hossenfelder and Palmer (2020).

[16] Aharonov et al. (2018).

[17] Aharonov et al. (2018).

[18] Busch (2009).

system becomes negligible. The price to be paid is that the first measurement becomes very poor. In other words, the observable becomes more and more smeared.[19]

Without having worked through the details, I think what is going on is this. The system starts in an entangled state, which, as we saw earlier in this section, fits the definition of strong emergence. Since the measurement is weak, the degree of entanglement after the measurement is only slightly reduced. So the system after the measurement is still entangled and so strongly emergent in exactly the same sense as entanglement in general. Not in a way that poses any problem for physics.

11.3 Room at the Bottom

Quantum physics is usually seen as indeterministic. Does this mean there is causal slack which higher levels could exploit? Let's start by remembering that quantum physics has two parts. The first describes the deterministic time evolution of quantum states. In general, these states will be superpositions of all the possibilities open to a system, recall the quantum coin of Sect. 2.2. The second describes a measurement process in which the superposition collapses to a single value. This measurement process is, under most interpretations, fundamentally indeterministic.

On the face of it, such indeterminism opens a channel for downward causation. You just need to describe a mechanism by which the higher level rigs the probabilities to achieve a particular outcome.[20] It certainly sounds like this is what Ellis intends when he writes that downward causation can happen

> …by statistical fluctuations and quantum uncertainty. Lower level physics is not determinate: random fluctuations and quantum indeterminism result in an ensemble of lower level states from which a preferred outcome is selected according to higher level selection criteria. Thus top-down selection leading to increased complexity is enabled by the randomness of lower level processes.[21]

---

[19] Busch et al. (2016), p. 244.

[20] A rare discussion of the implications of exploiting indeterminism, albeit in the context of free will, is in Liu (2009).

[21] Ellis (2016), p. 15.

Ellis along with Barbara Drossel propose a new interpretation of quantum physics[22] which they call contextual wavefunction collapse. The central idea is that quantum effects are limited in scale. For standard quantum physics, it is meaningful to talk of the wavefunction of a cat, a human or the universe as a whole. Instead, Ellis and Drossel argue that quantum physics has limited validity

> Even though the context of the quantum system can be described quantum mechanically at least for some of the degrees of freedom and up to certain length and time scales, the wider environment is a classical environment.[23]

They assert that this means collapse is an example of downward causation. One of the main attractions of this interpretation is that it doesn't require a separate collapse process. However, as far as I can tell the result of the collapse remains indeterminate. The paper does not include a description of how the higher level can rig the probabilities to choose a particular outcome. If I am right, then whatever the merits of Drossel and Ellis's proposal, it doesn't give a mechanism by which the higher level can affect the result of the collapse and so have causal effects.

Any proposal for such a mechanism would need to resolve two issues. The most evident is how to solve the staggeringly complicated computational problem of which collapses to choose to orchestrate a macroscopic effect. A deeper problem is that the molecular storm, the random motion of molecules, swamps quantum effects. Since the macroscopic world is largely robust to the molecular storm, it is likely to be wholly insensitive to individual quantum events. This is perhaps part of the reason why I failed to find any such proposals, at least outside the literature on free will.

A different approach comes from a 2017 paper[24] by David Yates. He claims that the geometry of the water molecule is an example of strong emergence. The fascinating properties of water, some of which are fundamental to cellular function and the existence of life, arise from its bond angle which was illustrated in Fig. 2.4. Yates asserts that, while quantum physics can calculate the properties of the molecule to a high precision, it can only do so taking the bond angle as given and that this implies the geometry of the water molecule has causal powers. So, in the sense of Wilson, it is strongly emergent.

---

[22] Drossel and Ellis (2018).
[23] Drossel and Ellis (2018).
[24] Yates (2017).

Yates states that the derivation "cannot be done".[25] But he uses words rather than quantum physics. And these words appear to be contradicted by two recent papers that claim to derive the bond angle from first principles.[26] One uses a computational method known as Density Functional Theory to calculate the energy of the water molecule for various bond angles and finds it is minimised at the observed angle. The other simulates the molecule on a quantum computer and derives the bond angle. Interestingly, Wilson gives a comprehensive rebuttal of Yates's argument from a philosophical perspective.[27]

11.4 Strong Computational Emergence

We saw in Sect. 5.8 that there are physical limits to computation. Mark Pexton applies this to the universe as a whole and argues that its resources are insufficient to compute everything in it. If this is the case this means that to do what it does, the universe itself must use some sort of higher level algorithm which Pexton asserts is analogous to the special sciences. For example, when a ball rolls down a slope the universe doesn't solve the quantum many-body problem, but instead uses good old Newtonian physics. This implies the special sciences are not arbitrary but have ontological significance since they are necessary for the universe to get on with its business. Pexton calls this strong computational emergence.[28] It's strong emergence in the sense that higher levels are not just approximations but have genuine causal powers.

He cites two sorts of evidence for his contention. The first is based on the combinatorial explosions which I already mentioned in Sect. 2.12. He takes the case of a protein made of a combination of 20 amino acids. A small protein might consist of a chain of 100 amino acids so there are 20^{100} possible such proteins. This number far exceeds the estimation of the computational capacity of the universe. How, then, do biological systems ever manage to find the correct protein? The second example relates to a class of mathematical problems called NP-complete[29] that take similarly implausible computational resources to solve then. He argues that some physical processes are themselves NP-complete but that the universe solves them effectively instantly.

---

[25] Yates (2017), p. 834.

[26] Respectively, Milovanović et al. (2020) and Xu et al. (2024).

[27] Wilson (2021), Sect. 4.3.

[28] Pexton (2019).

[29] These problems are a subset of the NP-hard problems discussed in Box 5.2.

Yet neither example bears closer examination. Proteins are selected by the directed search that is evolution. No biological process exists that requires running through all possible proteins. It's as simple as that. On the second example, a 2005 paper[30] by Scott Aaronson demolishes the idea that physical systems solve NP-complete problems. Instead, they settle into local optima rather than the global optimum which would represent the solution to the NP-complete problem. The article ends with the suggestion that the inability to solve NP-complete problems should be a constraint on physical theories.

11.5 Non-computability

Remember the discussion of computability in Sect. 5.4? I framed this in terms of the halting problem. In general, it is impossible to decide whether a particular Turing machine will stop or not. If it could be shown that a physical system was non-computable, this would imply that, even with a complete knowledge of the microphysics, there would be some properties that were impossible to derive. While this does not endow those properties with causal powers, so fails to be strong emergence in the sense of Wilson, it does imply that those properties are necessarily irreducible to physics. This would represent a challenge to physicalism so merits inclusion in this chapter.

There have been two recent demonstrations of the non-computability of physical systems. Both proceed by constructing a formal parallel between the system and a Turing machine. Then, since we know the halting problem applies to Turing machines, it must apply to the physical system i.e. the system has a property which cannot be computed.

The first paper,[31] from 2008, works with the Ising model which was the subject of Sect. 10.4. The parallel is established by showing that the Game of Life can be encoded in the Ising lattice. Then, since we know we can build Turing machines in the Game of Life, we can also encode them in the lattice. Since the halting problem applies to some Turing machines, so the configurations of the lattice which represent these Turing machines must be non-computable.

The second paper, from 2015, deals with the spectral gap,[32] the difference between the lowest and next-lowest energy levels of a material. They show that a Turing machine can be encoded in the quantum states of a material

---

[30] Aaronson (2005).

[31] Gu et al. (2009).

[32] Cubitt et al. (2015).

in such a way that the energy state depends on whether the Turing machine halts or not. This implies that the spectral gap is undecidable.

These are fascinating papers. But, inevitably since they are based on the halting problem, both only apply to infinite systems. It is not clear if there are any implications for finite physical systems.

You can approach the issue of non-computability from the other end. A process which can give answers to non-computable problems is conventionally known as an oracle. It's easy to imagine a black box that could solve the halting problem: give it any program, and the output tells you whether it will halt or not. You test it by giving it every program you can find, even ones designed to be tricksy, and it always gives you the right answer. While this might be evidence of something supernatural (imagine if you open the box and find it empty) it can never be evidence of non-computability. You can only test it with a finite number of finite programs so you can never know whether the oracle solves the halting problem in general. It may be that it just has an efficient algorithm which works in all the cases you feed it. Once more, the physical relevance of non-computability is not clear.

Does all this imply that non-computability is irrelevant for finite systems and hence the finite universe? Not necessarily. If physical quantities take continuous values, the infinity that matters would reside in these values rather than the size of the system. But are we sure that physical quantities are continuous? Einstein wrote

> One can give good reasons why reality cannot at all be represented by a continuous field. From the quantum phenomena it seems to follow with certainty that a finite system of finite energy can be completely described by a finite set of numbers (quantum numbers)[33]

Quantum physics provides some support for this idea because it states that some physical quantities are discrete. Are space and time similarly quantised? No one knows. Some theories of quantum gravity imply quantisation, but there is no consensus. Such quantisation would have empirical consequences but no experiment can ever definitively show its absence since the quanta could always be just a bit smaller than the accuracy of the experiment.

Just as computation requires physical resources, so does the storage of information. So a physical quantity with an infinite number of digits would require infinite resources. Of course, just as the universe may compute using a super-efficient algorithm, it may also have a super-efficient way of storing information. But if this is not the case, one solution is simply to argue that

---

[33] Einstein (2005), p. 99.

physical quantities only take values that do not require an infinite number of digits. In such a universe, everything would be computable.

11.6 Emergent Dualism

So far, everything I've discussed has assumed there is nothing in the universe apart from what is described by current or future physics. Emergent dualism is the idea that particular configuration of matter either brings something non-physical, let's call it spirit, into existence or attracts a pre-existing spirit. If this spirit is to be other than epiphenomenal, it needs at least some causal powers independent of its microphysical base and these will necessarily violate physical causal closure. There is an interesting debate over exactly how this happens and whether spirit must necessarily contradict conservation laws. But happen it must if spirit is to have causal powers: physics just does what it's told by spirit, in a way that, so far at least, avoids empirical measurement. In Chap. 13, I'll say a bit more about such possibilities and the complete absence of evidence for dualism.

11.7 Discussion

While in my opinion there is absolutely no evidence in favour of strong emergence, it would be perfectly reasonable to argue that this is because we haven't looked hard enough. Image there was a downward causation mechanism that only switches on when macroscopic samples are in a precise configuration. The high energy experiments that give the most precise measurements only involve a few particles at once so wouldn't pick it up. Condensed matter physicists may not have got round to working with the particular configuration.[34] This might lead you to conclude that the only response to strong emergence is agnosticism.

However, outside the free will debate, I am not aware of a single concrete suggestion of how downward causation could either exploit quantum indeterminism directly, by rigging the probabilities, or by somehow sidestepping causal closure. Even more striking is the complete lack of predictions that could help test ideas about strong emergence. If we are to be agnostic towards strong emergence, it is in same sense as we are agnostic to pixies in our garden.

---

[34] To avoid being picked up in equally high-precision astrophysical measurements, the hypothetical force would also need to fade away at large scales.

It is simply not worth devoting time to such things until there is evidence or testable predictions.

What would it mean for physicalism if evidence was found for strong emergence? Evidence for one of the mechanisms described in the first three sections would mean new physics, either directly or to explain the mechanism by which the high level rigs the quantum probabilities. This would extend physics and hence the scope of physicalism. On the other hand, if there were evidence for emergent dualism, we would have a clear sense of the limits of physicalism. The term strong emergence is redundant: it is either dualism or an extended physicalism.

11.8   Further Reading

For an interesting collection of papers related to downward causation, see Voosholz and Marcus (2022). For oracles and non-computability, see Edis and Boudry (2014). General discussions of emergent dualism can be found in Nida-Rümelin (2007) and Wong (2019). Explorations of how spirit might interact with physics are in Gibb (2010) and Cucu and Pitts (2019).

More suggestions for reading can be found at www.TheMaterialWorld.net.

References

Aaronson S (2005) Guest Column: NP-complete problems and physical reality. SIGACT News 36:30–52. https://doi.org/10.1145/1052796.1052804

Aharonov Y, Cohen E, Tollaksen J (2018) Completely top–down hierarchical structure in quantum mechanics. Proceedings of the National Academy of Sciences 115:11730–11735. https://doi.org/10.1073/pnas.1807554115

Aspect A, Grangier P, Roger G (1982) Experimental Realization of Einstein-Podolsky-Rosen-Bohm Gedankenexperiment : A New Violation of Bell's Inequalities. Phys Rev Lett 49:91–94. https://doi.org/10.1103/PhysRevLett.49.91

Busch P (2009) "No Information Without Disturbance": Quantum Limitations of Measurement. In: Shimony A, Myrvold WC, Christian J (eds) Quantum reality, relativistic causality, and closing the epistemic circle: essays in honour of Abner Shimony. Springer, Berlin

Busch P, Pekka L, Pellonpää J-P, Ylinen K (2016) Quantum measurement. Springer, Berlin

Clayton P (2006) Mind and emergence: from quantum to consciousness. Oxford University Press, Oxford

Cubitt TS, Perez-Garcia D, Wolf MM (2015) Undecidability of the spectral gap. Nature 528:207–211. https://doi.org/10.1038/nature16059

Cucu AC, Pitts JB (2019) How Dualists Should (Not) Respond to the Objection From Energy Conservation. Mind and Matter 17:95–121. https://www.mindma tter.de/resources/pdf/pittswww.pdf

Davies PCW (2012) The epigenome and top-down causation. Interface Focus 2:42–48. https://doi.org/10.1098/rsfs.2011.0070

Deacon TW (2012) Incomplete nature: how mind emerged from matter. W.W. Norton & Co, New York

Drossel B, Ellis G (2018) Contextual Wavefunction collapse: an integrated theory of quantum measurement. New J Phys 20:113025. https://doi.org/10.1088/1367-2630/aaecec

Edis T, Boudry M (2014) Beyond Physics? On the Prospects of Finding a Meaningful Oracle. Found Sci 19:403–422. https://doi.org/10.1007/s10699-014-9349-z

Einstein A (2005) The meaning of relativity. Princeton University Press, Princeton

Einstein A, Podolsky B, Rosen N (1935) Can Quantum-Mechanical Description of Physical Reality Be Considered Complete? Phys Rev 47:777–780. https://doi.org/10.1103/PhysRev.47.777

Ellis G (2016) How can physics underlie the mind? Springer, Berlin

Gibb S (2010) Closure Principles and the Laws of Conservation of Energy and Momentum. Dialectica 64:363–384. https://doi.org/10.1111/j.1746-8361.2010.01237.x

Gu M, Weedbrook C, Perales A, Nielsen MA (2009) More Really is Different. Physica D: Nonlinear Phenomena 238:835–839. https://doi.org/10.1016/j.physd.2008.12.016

Hawking SW, Penrose R (2000) The nature of space and time. Princeton University Press, Princeton, N.J

Hossenfelder S, Palmer T (2020) Rethinking Superdeterminism. Front Phys 8:139. https://doi.org/10.3389/fphy.2020.00139

Liu C (2009) How We May Be Free from Physics. Chinese Journal of Foreign Philosophy 20:99–148. https://philsci-archive.pitt.edu/3017/

McLaughlin BP (2008) The rise and fall of British emergentism. In: Bedau M, Humphreys P (eds) Emergence: contemporary readings in philosophy and science. MIT Press, Cambridge, Mass

Mcleish TCB (2017) Strong emergence and downward causation in biological physics. Philosophica 92:. https://doi.org/10.21825/philosophica.82113

Milovanović MR, Živković JM, Ninković DB, et al (2020) How flexible is the water molecule structure? Analysis of crystal structures and the potential energy surface. Phys Chem Chem Phys 22:4138–4143. https://doi.org/10.1039/C9CP07042G

Nida-Rümelin M (2007) Dualist Emergentism. In: McLaughlin BP, Cohen JD (eds) Contemporary debates in philosophy of mind. Blackwell Pub, Malden, Mass

Penrose R (1999) The emperor's new mind: concerning computers, minds and the laws of physics. Oxford University Press, Oxford

Pexton M (2019) Computational Emergence : Weak and strong. In: Gibb SC (ed) The Routledge handbook of emergence. Routledge, New York

Sailer T, Debierre V, Harman Z, et al (2022) Measurement of the bound-electron g-factor difference in coupled ions. Nature 606:479–483. https://doi.org/10.1038/s41586-022-04807-w

Sheldrake R (1995) A new science of life: the hypothesis of morphic resonance. Park Street Press, Rochester, Vt

Silberstein M, McGeever J (1999) The Search for Ontological Emergence. Philosophical Quarterly 49:201–214. https://doi.org/10.1111/1467-9213.00136

Touboul P, Métris G, Rodrigues M, et al (2022) M I C R O S C O P E Mission: Final Results of the Test of the Equivalence Principle. Phys Rev Lett 129:121102. https://doi.org/10.1103/PhysRevLett.129.121102

Voosholz J, Marcus G (2022) Top-Down Causation and Emergence. Springer Nature, S.l.

Vopson MM (2019) The mass-energy-information equivalence principle. AIP Advances 9:095206. https://doi.org/10.1063/1.5123794

Vopson MM (2022) Experimental protocol for testing the mass–energy–information equivalence principle. AIP Advances 12:035311. https://doi.org/10.1063/5.0087175

Walker SI, Cisneros L, Davies PCW (2012) Evolutionary Transitions and Top-Down Causation. https://doi.org/10.48550/arXiv.1207.4808

Wilson JM (2021) Metaphysical Emergence. Oxford University Press, Oxford

Wong HY (2019) Emergent dualism in the philosophy of mind. In: Gibb SC (ed) The Routledge handbook of emergence. Routledge, New York

Xu F, Yang F, Wei C, et al (2024) Quantum simulation of water-molecule bond angles using an NMR quantum computer. Phys Rev A 109:042618. https://doi.org/10.1103/PhysRevA.109.042618

Yates D (2017) Demystifying Emergence. Ergo, an Open Access Journal of Philosophy 3:. https://doi.org/10.3998/ergo.12405314.0003.031

12

Emergence: An Assessment

Summary This chapter presents a brief synopsis of the arguments in Part II that show emergence is an illusion.

We've seen that weak emergence is such a weak concept that it applies to everything from quarks to the whole universe. We've seen there are no examples of strong emergence. So emergence either applies to everything or to nothing. We think the concept tells us something about the world but this an illusion. This chapter briefly summarises the arguments that lead to this conclusion.

12.1 More is Different

What it is: wholes have properties distinct from those of their parts.

Why it applies to everything: it is the case for all composite bodies.

Why it is not a challenge to reductive physicalism: more is different because parts interact and such interactions have always been at the heart of physics.

L. Graham, *Physics Fixes All the Facts*, The Frontiers Collection, https://doi.org/10.1007/978-3-031-69288-8_12

12.2   Simulation Emergence

What it is: weakly emergent phenomena are those for which no closed-form solution exists and so can only be modelled by simulation.

Why it applies to everything: exact closed-form solutions only exist for idealised systems.

Why it is not a challenge to reductive physicalism: it doesn't claim to be since it is about models, not the world.

12.3   Multiple Realization Emergence

What it is: high-level properties somehow float free of physics since they can be realized by systems with different microphysical structures.

Why it applies to everything: everything which is not fundamental physics is a high-level property and so emergent in this sense.

Why it is not a challenge to reductive physicalism: every physical system is distinct. MR emergence captures the simplifications required by the nature of human thought.

12.4   Contextual Emergence

What it is: physics only sets the necessary conditions for the behaviour of a system; the sufficient conditions are given by the context.

Why it applies to everything: only the most idealised systems are independent of their context.

Why it is not a challenge to reductive physicalism: context does not appear by magic but is itself a physical system. Then all contextual emergence says is that one physical system, the experiment, interacts with another, the context.

12.5 Strong Emergence

What it is: higher levels have causal effects on their microphysical base, either by violating physical causal closure or by sidestepping it.

Why it doesn't apply to anything: although logically coherent, there are neither convincing theoretical models of strong emergence nor empirical evidence.

Why it wouldn't be a challenge to reductive physicalism: if evidence of strong emergence were found and could be explained by new physics, this would extend the scope of physicalism.

Why it would be a challenge to reductive physicalism: if evidence of strong emergence were found and the only explanation was dualist, this would put a limit on the scope of physicalism.

12.6 And so?

Our cognitive limitations mean we cannot do without high level concepts. But we can do without the concept of emergence. Chapter 14 will show what we can replace it with. Before that, we need a general framework with which to approach complex systems and indeed physical systems in general. That is the subject of the next chapter.

Part III

The Demon's Perspective

13

Austere Physicalism

Summary Of the six positions presented in Chap. 3, we have so far ruled out weak and strong emergence so are left with dualism and the three types of physicalism. This chapter will argue that the only coherent physicalist position is eliminativist. This means it must be an austere physicalism in which the only real entities are those described by fundamental physics. Everything else, everything that is not physics, is an illusion. After stating the position and giving some philosophical background, the bulk of the chapter considers various objections and concludes that none of them are convincing.

Emergence is an illusion. If we want to avoid the supernatural, what is left? Part III argues that the only choice is an austere physicalism. This chapter presents the physicalist position and investigates its consequences. The next chapter uses it to unpack the concept of emergence. Then Chap. 15 returns to the examples of Chap. 2 and shows how they can be understood in a physicalist framework. Finally, Chap. 16 will sketch the implications of physicalism for free will and consciousness.

Chapter 3 described six metaphysical positions: dualism, epiphenomenalism, reductionism, eliminativism, weak emergence and strong emergence. This chapter argues that we are left with a choice of two: dualism or eliminativism. What happened to the other four? We've seen that strong emergence would either mean new physics or dualism, but that there is not a shred of evidence for it. And that the various formulations of weak emergence tell us nothing useful about the world since they can be applied to everything.

L. Graham, *Physics Fixes All the Facts*, The Frontiers Collection,
https://doi.org/10.1007/978-3-031-69288-8_13

Of the three physicalist positions, epiphenomenalism makes no sense to me and I'm going to ignore it. This leaves eliminativism and reductionism. You might say that, given the choice, you'd stump for reductionism. At least it allows you to preserve your intuitions about the reality of the things that compose your world. But I am going to argue that it is an unstable position. A reductionist cannot avoid being an eliminativist. You can't stop everything slipping down to the lowest level. You can't avoid seeing that only the lowest level is real. This means that the only possible physicalism is an austere physicalism. It is either that or a belief in the supernatural.

13.1   Everything is Physics

Everything is fully described by fundamental physics or whatever future theory may supersede it. Physics explains the parts, the whole and the relation between them. Due to the non-local nature of quantum physics, wholes can influence parts as well as parts influencing wholes. Other ways of describing the world are artefacts of our cognitive limitations. This includes the special sciences, commonsense models of the world, ordinary objects, ourselves and other creatures. They may be useful approximations. They may be unavoidable given our cognitive structure. But they are illusions in the sense they have no causal power.

This means that almost everything we think we know about the world is wrong. The sky isn't blue. Nothing is blue. There are no colours. You are not holding a book. There are no books. Instead, there are quantum fields arranged blue-wise or book-wise. These interact with other quantum fields arranged person-wise and so reconfigure them into a state that corresponds to seeing a blue sky or holding a book. There are no objects above the lowest level. Indeed, there are no levels. There are only quantum fields arranged in various ways.

If quantum physics is beyond our imaginative understanding, its implications for ourselves and the universe as a whole will necessarily make no sense to us. But let me try engage your imaginative understanding by telling a story that I hope will make things more plausible. A few minutes after the big bang, the universe was a mix of subatomic particles at a temperature of around 10^{10}K. The particles were distributed approximately uniformly. Yet 14 billion or so years later the universe is full of complex structures at every scale, from biological nanomachines up to clusters of galaxies.

How did we get from there to here? I gave an outline in Sect. 9.4, but the important point here is that, if physicalism is true, each step happened according to the laws of physics. At every step, the dynamics of every particle in the universe evolved according to quantum physics and general relativity.[1]

If you accept that fundamental physics explains the development of the universe, you are forced to accept that it explains the state of the universe at any point in this evolution. And this means it describes every detail of everything in the universe. This includes here and now. Everything is physics. We might never be able to follow every detail of the causal chain, I'll have more to say about this in the next chapter, but physical causal closure means that the chain exists.

The only alternative is to assert that somewhere in the causal chain leading from the early universe to the present, something non-physical slipped in. This would be dualism.

Austere physicalism is a radical position. Objections to it can be divided into three broad types: that it clashes with our intuitive understanding of the world, that it clashes with the practice of science or that its consequences are unacceptable. Before engaging with such objections, let's address the question of where the process of elimination ends.

13.2 Slicing the Blobject

Continuing the process of elimination means that there is only one object. Terry Horgan and Matjaž Potrč call it the *blobject*:

> There is really just one concrete particular, namely, the whole universe (the blobject).
> The blobject has enormous spatiotemporal structural complexity and enormous local variability—even though it does not have any genuine parts.[2]

The blobject was introduced as a philosophical idea. However, we can put it on physical foundations. Quantum physics implies that everything in the universe forms one entangled quantum state. This is an almost trivial consequence of the big bang. If all the matter in the universe started out in an entangled state it must remain in one.[3] This means it is described by a single wave function. Putting some region of the universe, like the quantum coin of

---

[1] The point does not depend on this evolution being deterministic. If quantum physics has an indeterministic component, we just need to say that the universe evolved according to physical rules conditional on this random element.

[2] Horgan and Potrč (2008), p. 3.

[3] This doesn't preclude the degree of entanglement increasing over time, indeed a 2024 paper, Al-Khalili and Chen (2024), argues that this may give another interpretation of the arrow of time.

Sect. 2.2, in a coherent state and then watching it decohere is a purely local effect. This view is known as quantum monism.

Of course, such a position is dependent on which interpretation of quantum physics one chooses. But that is not the issue here. To think more about the implications of the blobject, let's take what might be called the demon's perspective. Pierre-Simon Laplace used the thought experiment of a superbeing, conventionally known as a demon:

> Given for one instant an intelligence which could comprehend all the forces by which nature is animated and the respective situation of the beings who compose It—an intelligence sufficiently vast to submit these data to analysis— it would embrace in the same formula the movements of the greatest bodies of the universe and those of the lightest atom; for it, nothing would be uncertain and the future, as the past, would be present to its eyes.[4]

Laplace's target was determinism, but let's train the demon's gaze on the blobject. Before you object that computational limits apply to demons too, I intend this as no more than a metaphor. Our demon would be able to see the blobject in its entirety, the single wave function (or whatever form the true description takes) of the universe as a whole. If the multiverse interpretation of quantum physics is correct, the demon would see every detail of every universe simultaneously. It would be able to run this wave function back and forward in time, from the moment of the big bang into the eternity of the big freeze.

The demon could effortlessly slice up the blobject in any possible way, across dimensions and across timescales. Some subset of these would correspond to things in our scientific picture of the world. The demon could appreciate the delicate dance of microstates in a gas, pausing at each tick of the Planck clock (or whatever is the minimal increment of time, if there is one) to watch the slow evolution of its wave function. It could watch the last atom fall into a black hole then trace its path back to its formation as the early universe cooled.

The demon could work out the special sciences and all their laws and see why particular physical systems (humans) at a particular stage in their development have a particular set of special sciences. And at the same time see why other physical systems (all the other ways that intelligence can be instantiated) would have different special sciences.

A smaller subset of the demon's perspective would contain the things in our commonsense picture of the world. The demon would see that some

---

[4] Laplace (1995).

physical systems when close to other physical systems have states which correspond to "seeing blue" and "holding a book". It could run back the development of these systems to the first bacterium that represented the external world by some internal state. And continue further back to the first set of self-replicating molecules. It would see all of these processes in terms of energy minimisation or entropy maximisation, or other physics unknown to us. Perhaps other physical systems exist at other places and times with vastly greater complexity than humans. The demon would see them all in exactly the same way, indifferent, as the Great Programmer of Sect. 5.10 is indifferent, to whether they are configured in ways that represent life or consciousness.[5] Speculating about the rest of what the demon sees, the part that lies outside our scientific image, is no better than science fiction. The blobject can be sliced in endless different ways from the smallest scale to the largest.

Will we ever achieve something like the demon's perspective? If the Church-Turing-Deutsch principle is true, the universe is comprehensible. If we can build a universal quantum computer, every system can be simulated from the bottom up subject only to the limits of the size of the quantum computer. This would mean there are no limits to physicalism. Or there may be general limits to the system of distributed cognition that is science, I will return to this question at the end of the chapter, and we may one day run up against them. But until this happens, science and the physics that underlies it will continue to explain more and more of the world. Austere physicalism is profoundly optimistic.

13.3   The Philosophical Background

The term *eliminative materialism* was first used in a 1968 article by James Cornman,[6] though similar ideas can be found in the writings of the British Emergentists in the 1920s.[7] Subsequently, the use of the term has split into two threads.

The first aims eliminative materialism squarely at commonsense psychology. One of the best known contributions is Paul Churchland's

---

[5] There is a close parallel between the blobject and the Great Programmer's view of our universe as one very long integer. Our special science view is one way of compressing this integer. A demon would see many others.

[6] Cornman (1968).

[7] For example, Broad (1925), p. 610.

"Eliminative Materialism and the Propositional Attitudes". Here is its first paragraph:

> Eliminative materialism is the theory that our common-sense conception of psychological phenomena constitutes a radically false theory, a theory so fundamentally defective that both the principles and the ontology of the theory will eventually be displaced, rather than smoothly reduced, by completed neuroscience.[8]

This is strong stuff. Churchland's body of work forms the core of the eliminativist argument against commonsense psychology. However, as far as the special sciences go, Churchland seems to remain a reductionist.

The second strand takes the complementary approach: being eliminativist with respect to ordinary objects (and implicitly to the special sciences) while asserting that some classes of living things are different.[9] This is Peter van Inwagen:

> The y composed by the x's exists if and only if the activity of the x's constitutes the individual life of a concrete biological organism[10]

And this Trenton Merricks:

> We humans—in virtue of causing things by having conscious mental properties—are causally non-redundant[11]

I think van Inwagen's argument is easy to dispose of by appealing to continuity. There is no sharp cut off between non-living and living matter. If there appears to be one, that is only because we are thinking of the highly developed forms of living matter we see around us. Instead, if we tell a story of how life started, there is a smooth transition from non-life to life.[12] So all that's left is to claim that matter organised as living things somewhere undergoes a step change as the magic flame of life is lit. This would be vitalism, one form of dualism. Merricks argument evokes consciousness which puts it, at least until the final chapter, outside the scope of this discussion.

My approach combines these two threads. I'm not calling it eliminative materialism for two reasons. The first is that the term is mostly associated

---

[8] Churchland (1981).
[9] For a comparison of the two positions, see Van Inwagen (2014), p. 6.
[10] Van Inwagen (1995), p. 82.
[11] Merricks (2006), p. 114.
[12] See Graham (2023).

with just the first thread. The second is that I've settled on the term physicalism (Sect. 3.13). Since elimination applies equally to ordinary objects, special sciences and commonsense psychology, we are close to the Austere Realism of Horgan and Potrč. Hence Austere Physicalism, as a link and tribute to their work.

13.4 Objection: It's Blatant Nonsense

I wonder why any sane reader would give the time of day to such a preposterous claim… physics can tell us about an atom of metal accelerating through the air, but can it say anything about the swing of the executioner's axe, or whether capital punishment is a good thing?[13]

This was the response of one of my friends when I pitched austere physicalism to them. And it seems quite a common one. Physics is fine in its domain. But step into the complex realm of human affairs and it is just plain silly to think it has anything to say. To explain things in the world, you need a completely different set of concepts. You might recognise this objection as a version of the Churchill's Nose example of Sect. 2.15. I'll postpone a thorough discussion of this until Chap. 15, but for the moment, let me respond in two ways.

First, with a question. If these things (history, morality) are not physics, what are they? There is a complex causal chain leading from the early universe to the executioner's axe. Where in this chain does something non-physical slip in? And, in the absence of a meaningful sense of emergence, what exactly is this non-physical thing?

Second, with an analogy. In Sect. 9.3, I discussed Bishop and Ellis's argument that it is absurd to imagine that the distribution of matter some minute fraction of a second after the big bang could somehow encode these words that I am writing and everything around me. My response to this was that the nature of quantum physics means information is conserved and the information content of the universe is constant. Yes, it is impossible to accept this. But we know how limited is our imaginative understanding of the world. We need to put it aside and accept the representational understanding given by physics.

---

[13] Honor Klein, personal communication.

Some philosophers go further than my friend and argue that eliminativism is self-contradictory.[14] Briefly, the argument goes that eliminativism asserts that there is no meaning in the world so if it is true cannot itself be meaningful. Alex Rosenberg gives a convincing rebuttal in a 1991 paper.[15]

13.5   Objection: It's not How We Think

Of course it's not. When we see a blue sky or hold a book, it's not that we're making a mistake. Our sensory system is configured in such a way that the only way we can see the world is in terms of colours and objects. We have no choice but to use this *commonsense ontology* which is encoded in the brain and the result of a long process of evolution.

Wilfred Sellars refers to the objects in the commonsense ontology as the *manifest image*[16] This is Daniel Dennett's description:

These are the things we use in our daily lives to anchor our interactions and conversations, and, to a rough approximation, for every noun in our everyday speech, there is a kind of thing it refers to. That's the sense in which the 'image' is 'manifest': it is obvious to all, and everybody knows that it is obvious to all, and everybody knows that, too.[17]

The manifest image is the view from the cave. Commonsense ontology tells us what there is in this world. Commonsense physics, statistics and psychology tell us how these things behave and interact. We have imaginative understanding of all these things. In fact, to imaginatively understand something is to fit it into the manifest image.

Sellars contrasts the manifest image with the *scientific image*[18] We've seen that science is a system of distributed cognition, a system that allow us to see the world outside the cave. For an austere physicalist, this consists of fundamental physics and nothing else. We have only representational understanding of this. Our minds are too limited for imaginative understanding.

---

[14] This argument is in Broad (1925), p. 611.

[15] Rosenberg (1991).

[16] Sellars (1992), p. 6.

[17] Dennett (2017), Chap. 4.

[18] You can find a similar idea in the work of Galileo, see Drake (1960), p. 309.

Using terms from the manifest image, we might say "The sun has just moved behind the house". But we wouldn't dream of using this as evidence to deny Copernicus's insight that the sun doesn't move. Science can help us understand why it seems to us that the sun has moved, how such a perception is the result of our evolutionary history and how it plays a useful role in discourse. In a similar way, we can't help thinking or saying things like "That's a lovely blue sky" or "I'm reading a book". But such commonsense usage has little to do with physical reality which only contains quantum fields. The statement "I'm reading a book" can be no more a denial of austere physicalism than "The sun has just moved behind the house" is a denial of heliocentrism.

While the manifest image is a poor guide to the scientific image, the scientific image includes an explanation of the manifest image. If we want to know why we have a particular set of models, we need to study the evolutionary process that led to modern humans. In Sect. 4.1, I gave a sketch of such an explanation, of how our cognitive capacities are consequences of evolutionary constraints. Commonsense models represent good-enough heuristics which helped us survive and thrive in our ancestral environment. They are just one among endless possible ways to describe the world. They are about utility, not reality. There was never a selective advantage to understanding the microscopic or cosmological details of the world. The extraordinary thing is that we manage to see out of the cave at all.

What more can we say about the manifest image? Daniel Dennett describes the simplifications involved as various stances.[19] He gives the example of an alarm clock or a goldfish. Both are complex systems requiring a huge amount of analysis to understand how they work. But this isn't much use for practical purposes, and indeed wasn't available for most of history. Instead we adopt a *design stance*. We assume that the object has been designed for a particular function. This allows us to predict its behaviour in a much simpler way. The alarm clock will ring when the time reaches a particular value. The goldfish will swim in water but flounder in air. There are weaknesses to the design stance (we may misinterpret the design, or maybe the object won't work as intended) but these are compensated by the vast increase in ease of prediction.

However, for the goldfish, a further simplification arises from treating it as an agent, which Dennett calls the *intentional stance*. This involves using the tools of commonsense psychology, ascribing beliefs, desires and some level of

---

[19] Dennett (2013), Chap. 18.

rationality to the system. The goldfish will swim to the surface when food is scattered because it is hungry. It will interact with other goldfish because it is playful, and so on. As for the goldfish, so for any living system from a bacterium upwards.

Living systems are complicated. As a simpler example of the workings of the intentional stance, take a four-wheeled machine equipped with a motor.[20] It has a light sensor on each side and the stronger the signal from the light sensor the faster the wheels on that side turn making it turn away from the light. Then if we put it on a flat surface lit in a non-uniform way it will tend to find the darkest part and turn in circles around it. It is much easier to say that the machine moves in order to find a dark spot rather than dig down into the messy details of how it is put together. But the machine has no more intention or purpose than a stone falling under gravity. Purpose is an illusion, something humans project onto physical systems and not something that exists in the world.

The design and the intentional stances ignore all the internal details of their subjects. Indeed, that is one of their great merits. Instead of having to worry about endless complexity, they provide a fast, frugal and generally reliable way of predicting behaviour. They are useful approximations that are wired into our commonsense model of the world.

Both stances are rich fodder for the mind projection fallacy. The design argument leads to the idea of a designer and when applied to the universe this gives the Teleological Argument for the existence of a deity. The intentional stance results in an enchanted world full of spirits and gods. Chapter 16 will argue that many of our illusions about mental causation come from applying the intentional stance to our own actions.

Commonsense notions are extremely useful. Indeed, we cannot do without them. But they have the same relation to physics as does the statement "The sun is moving behind the house". We cannot read off features of the world from features of our language.[21]

---

[20] This example is adapted from Braitenberg (2004).

[21] Thomasson (2010), p. 180 appears to claim that instead language is what allows the possibility for the existence of ordinary objects: "that our singular and general nominative terms have a basic conceptual content in the form of frame-level conditions of application and coapplication collectively established by competent speakers.".

13.6   Objection: Illusions Are All We Have

Try to live your life according to fundamental physics. You'll starve before you manage to get out of bed. In our daily lives, the manifest image is all we have. An austere physicalist perspective makes it a fascinating object of study. How do we manage to construct a world full of ordinary objects having apparent causal power? How do we give meaning and beauty to quantum fields?

Less romantically, we can investigate the origin of our commonsense notions in our evolutionary history. We can investigate why our common-sense models work so well in so many situations. We can marvel at how the scientific image starts from and then transcends the manifest image. We can come up with a hierarchy of concepts, ranking them according to how well they approximate physics. Such investigations integrate the manifest image within physicalism.

13.7   Objection: It's not How Scientists Think

The vast majority of scientists don't have much to do with fundamental physics. Instead they use the terms of whatever special science they work in. Doesn't this prove the limited domain of physics?

I addressed this argument during the discussion of MR emergence in Sect. 8.6. The central point was that the structure of the special sciences is a continuously evolving function of human capabilities and interests. Remember Wikipedia's 60 sciences beginning with 'a'? The special sciences and the idea of levels are no more than useful but arbitrary ways of slicing up the un-layered blobject.

Carl Gillet calls it a scientifically manifest image.[22] Or we can think of it as a special science stance. It includes things like atoms, molecules, viruses, cells, planets and so on. Just like the design and intentional stances it is a useful, perhaps necessary approximation.

---

[22] Gillett (2018), p. 13.

Box 13.1 Alien special science

If the structure of the special sciences is a function of our cognitive limitations, would a different set of limitations lead to a different set of special sciences? Early signs that AI might provide insight into this question came from reactions to DeepMind's Go playing system, AlphaGo. Experienced players describe the moves as "amazing, strange, alien… from an alternate dimension…downright incomprehensible".[23] AI is free from many of the constraints that limited the evolution of human cognition (Sect. 4.1). Most notably, the constraint on computational resources is weaker so AIs can look at a problem as a whole rather than having to split it up into steps.

A fascinating hint of how technology could demonstrate the arbitrariness of the special sciences was given in a 2022 paper.[24] This described an experiment of using an AI to analyse videos of physical phenomena (including pendulums, lava lamps and fire) and build an internal model of them. The authors analysed the variables that appeared in the model and found that many of them are different from those that appear in standard models and some make no intuitive sense at all. If other groups find similar results, the consequences will be profound since the variables we are interested in have a constitutive role in the special sciences we develop.

> Since machines can see the world in colors that we cannot see, hear the world in frequencies we cannot hear, and experience the world using senses we do not have, machines may help intuit new kinds of fundamental variables that we cannot imagine.[25]

Philosophers often treat the special sciences as objective. I have already cited Fodor on natural kinds. As another example, Wilson claims that special science entities are causally and ontologically autonomous from the underlying physics. As evidence she cites distinctive special science features, taxonomy and laws:

> …that special-science laws are seemingly distinctive and seemingly causal provides prima facie support for special-science entities' being causally autonomous—that is, distinctively efficacious—with respect to their underlying micro-configurations.[26]

The repetition of the word "seemingly" is significant. It seems to me that Wilson is making the same mistake as Fodor, describing the objects in the scientifically manifest image then projecting them on to the world.

---

[23] Chan (2017).

[24] Chen et al. (2022).

[25] Chen et al. (2022).

[26] Wilson (2021), p. 4.

The practice of science tell us about the nature of our cognition, not the nature of the world. Here's Philip Kitcher's eliminativist description of how to understand the scientific process:

> At the most fine-grained level, scientific change involves modifications of the cognitive states of limited biological systems. What are the characteristics of these systems? What kinds of cognitive states can they be in? What are their limitations? What types of transitions among their states are possible? What types are debarred? What kinds of goals and interests do these systems have?[27]

13.8   Objection: It's not How We Experience Ourselves or Others

You know perfectly well that you are more than a fluctuation in a quantum field, don't you? You know this in the same way you know a whole heap of other things: the way your visual system gives you an accurate representation of the world outside; where your body starts and ends; your sense you can freely choose your actions.

But these are all illusions. Our visual picture of the world is stitched together from limited information and can be tricked in endless ways. The rubber hand illusion shows how easy it is to extend body ownership to inanimate objects. Oliver Sacks popularised the inordinate number of ways the brain can go wrong and our apparently coherent picture of the world can fragment. Given this, it is naïve to think that our model of ourselves is accurate.

Commonsense psychology, using the design and intentional stances, allows us to make good-enough predictions of the behaviour of the overwhelmingly complex systems that are other people. In this sense, it is a powerful and successful theory. But it's no more generally valid than any other part of the manifest image.

Commonsense physics is a poor guide to how objects behave. Introspection, based on commonsense psychology, is an even worse guide to how the brain works. We probably can't avoid talking in terms of beliefs, desires, choices etc., though in Sect. 16.1, I will give a thought experiment to show how technology might change this.

---

[27] Kitcher (1995), p. 59.

13.9 Objection: I'd Rather be a Reductionist

Reductionism preserves the reality of high-level features. It seems to let you keep your intuitions about the reality of the things that compose your world. There's an immediate problem with this. The most convincing definition of reality is having casual power (Box 3.1). Yet if you are a reductionist, you accept physical causal closure. If higher levels have no causal power, they cannot be real. We are back to eliminativism.

As another way of looking at this, let me repeat the passage from Jaegwon Kim that I quoted in Sect. 3.13:

> There is an honest difference between elimination and conservative reduction. Phlogiston was eliminated, not reduced; temperature and heat were reduced, not eliminated. Witches were eliminated, not reduced; the gene has been reduced, not eliminated.[28]

This seems clear enough, until we realise that it is no more than a repetition of the discussion of the special sciences. Yes, terms like temperature, heat and the gene are useful. They earn their keep by playing a role in high level explanations. Here's Richard Rorty:

> Why does the realization that nothing would be lost by the dropping of 'table' from our vocabulary still leave us with the conviction that there are tables, whereas the same realization about demons leave us with the conviction that there are no demons? I suggest that the only answer to this question which will stand examination is that although we could in principle drop 'table', it would be monstrously inconvenient to do so…[29]

Pre-quantum chemistry is better than no chemistry at all. Phlogiston is better than no theory of heat. When I navigate on a hike, the assumption that the earth is flat works just fine. If you try, you can probably find a reason why explaining mental illness by demons is better than having no explanation at all. All Kim is saying is that given our cognitive structure we cannot do without high level concepts.

---

[28] Kim (2007), p. 160.
[29] Rorty (1965).

13.10 Objection: Stop Telling Scientists What to Do

Opponents of reductionism (and, a fortiori, of eliminativism) often imply that reductionists think all scientists should be doing quantum field theory.

I've never come across anyone who says this and my sample includes several less than sober high energy physicists. Indeed, Steven Weinberg (Nobel Prize for Physics, 1979), often portrayed as an arch-reductionist, says the opposite:

> ...reductionism is not a guideline for research programs, but an attitude toward nature itself. It is nothing more or less than the perception that scientific principles are the way they are because of deeper scientific principles (and, in some cases, historical accidents) and that all these principles can be traced to one simple connected set of laws.[30]

The objection is yet another instance of arguments being directed against a straw man (Sect. 6.5). It's perfectly possible to be an austere physicalist and see fascinating questions everywhere. In the face of our cognitive limitations and the richness of the world, the only methodology that makes sense is pluralism.

13.11 Objection: Austere Explanations Are not Explanations

Max Born describes an exchange between his wife and Einstein:

> She once asked Einstein: 'Well then, do you believe that it will be possible to depict simply everything in a scientific manner?' 'Yes,' said Einstein, that is conceivable, but it would be no use. It would be a picture with inadequate means, just as if a Beethoven symphony were presented as a graph of air pressure.[31]

I take this to mean that the domain of scientific explanations doesn't extend to subjective experience. We may know everything that can be measured about a Beethoven symphony, but this doesn't tell us what it would be like to hear it. Let me postpone my response until the discussion of consciousness in Chap. 16.

---

[30] Weinberg (1992).
[31] Born (1989), p. 158.

13.12   Objection: Reductionists Are Nasty People. Austere Physicalists Must Be Worse

…Nowadays (almost) no one likes reductionism.[32]

Those on the political right don't like reductionism.

Theologians and others, including, of course, public intellectuals on the political Right, will find a complete physio-chemical understanding of humanity threatening to human dignity, individual responsibility, and divine agency.[33]

Those on the political left don't like reductionism:

…radical scientists take [reductionism] to be characteristic of bourgeois science and partly to blame for the inappropriate technologies. Reductionism reflects the bourgeois interest in centralized control.[34]

Feminists don't like reductionism:

…a feminist science is one whose theories encode a particular world view, characterized by complexity, interaction, and wholism.[35]

And some physicists don't like reductionism. Ellis's book contains extended discussion of its baleful effects in areas including health care, literacy teaching and psychotherapy.[36] Robert Laughlin (Nobel Prize for Physics 1998) claims:

Nuclear weapons are, unfortunately, the most sensational engineering contribution of physics, something that catapulted the discipline to prominence in the 1950s and has colored it indelibly ever since. This coloring is inherently reductionist.[37]

This sounds bad. And if reductionism is bad, austere physicalism must be worse.

---

[32] Tahko (2021), p. 1.

[33] Rosenberg (2006), p. 9.

[34] Longino (2019), p. 194.

[35] Longino (2019), p. 187.

[36] Ellis (2016), sec. 7.3.

[37] Laughlin (2005), p. 100.

I don't have much to say in response, apart from a general discomfort about political beliefs imposing constraints on science. Of these criticisms, only the feminist one strikes me as interesting. Following on from the passage I quoted, Longino goes on to argue against this view maintaining, along with many feminists, that it is important not to conflate feminism with feminine and feminist science shouldn't be restricted to a particular world view.

Reductionism, or the straw man that regularly takes its place, is also criticized for being arrogant. But it seems to me austere physicalism is humble. Humble in the way it accepts and understands our limitations. Humble in the way it is wary of projecting these limitations onto the universe.

13.13 Objection: Horrendous Consequences

Austere physicalism seems to doom everything that makes life worth living, sucking beauty and meaning out of the world. If consciousness itself is eliminated, with it too goes the central wonder of human existence, the mysterious flame that lights us up from the inside. We would be left with a disenchanted wasteland in which no one would want to live.

My uncharitable response would be that the universe does not care a jot about our discomfort. In fact, the universe doesn't care about anything. If I am feeling more generous, I would argue that these fears are overstated. The elimination of consciousness, if it ever happens, would be a theoretical process which would affect our subjective experience not one iota. It's a bit like saying progress in astrophysics makes the night sky less awe inspiring, or knowing optics makes a rainbow less beautiful. I would argue the opposite: that the more we understand something, the more wonderful it seems. The night sky or a rainbow are even more astonishing when you know the science behind them. Neuroscientific insights allow me to marvel over how my brain manages to stitch sensory fragments together to make what seems to be a coherent picture of the world around me. John Searle asks:

> How can we square [a] self-conception of ourselves as mindful, meaning-creating, free, rational, etc., agents with a universe that consists entirely of mindless, meaningless, unfree, nonrational, brute physical particles?[38]

I think this can be answered in a way that only enhances human experience. But doing so is the work of a different book.

---

[38] Searle (2007).

13.14 Evidence

Evidence for dualism would lead us to reject physicalism or at least limit its scope. As far as I know, there is none. This is all the more extraordinary given it is easy to imagine what such evidence would look like. Andrew Melnyk writes:

> …we might have discovered that the human skull, which evidently contains at least the proximate causes of human behavior (since it is where muscle-stimulating motor neurons come from), was empty, or full of blood, or (less implausibly) that it housed an organ of relatively modest complexity; and had we discovered any of these things, the postulation of a nonphysical mind would surely have been irresistible[39]

Instead, the skull contain the most complex system we've ever come across and one that seems built for cognition and information processing.

Less dramatic forms of dualism should be testable. If spirit affects the brain, it should also be able to affect other systems such as a measurement device. Or imagine if cognitive correlates of consciousness were identified and then found all the way down to bacteria. Physics might be able to incorporate such evidence with new concepts and laws. Elimination means finding the minimal set of entities needed to explain the universe. If this meant that we had to extend physics to include, say, a field responsible for producing consciousness, then as long as this field was susceptible to scientific treatment so be it. Otherwise, physicalism would have to be rejected.

Turning to examples that don't involve mental phenomena, there's a science fiction story[40] in which malevolent aliens decide to weaken human society by making the results of scientific experiments become random. The result is nervous breakdowns among scientists (who in this story are surprisingly fragile characters) and the end of scientific and technical progress. Imagine that reproducibility broke down in this way, with experiments that used to be well-behaved suddenly giving unpredictable results.

In a 2014 paper,[41] Taner Edis and Maarten Boudry give these lovely examples. What if researchers found a sequence of DNA shared by all life and, after repeated checking by every means possible, it is found to read "© Yahweh 4004 BC". Or if multiple studies of people visiting Lourdes showed that a

---

[39] Melnyk (2020), p. 186.
[40] Liu (2014).
[41] Edis and Boudry (2014).

statistically significant proportion were cured but only if they were devout Catholics.

Or if ghosts were often seen. Or I received, preferably direct to my mind, consistently accurate predictions of future events. Or if the dead rose and spoke to the living. It is easy to think of events which would force a reasonable person to conclude that physicalism was false.

13.15 Demons Just Want to Have Fun

The greatest trick the devil ever pulled was convincing the world he didn't exist[42]

Physicalist and dualist positions seem radically opposed. But physicalism is a much more modest doctrine than it might seem. All physics does is explain the phenomena. This leave lots of scope for dualism. To explore this, let's again evoke demons. I'll introduce you to three types: a scientific demon, who plays by the rules; a mischievous demon, who circumvents them and a deceiving demon, who writes the rules to hide its actions.

Scientific demons. We saw in Sect. 11.3 that quantum indeterminism could provide causal slack for downward causation to act. In a similar way, if nothing physical determines the state to which a wave function collapses, then there is a space for a demon to do so. Could this have macroscopic effects? The problem is the same as for strong emergence: the molecular storm swamps quantum effects. This means that a scientific demon would need to orchestrate many quantum collapses to have an effect and, to remain scientific, would need to do so in a way that was undetectable.

However, the molecular storm would also give a scientific demon scope to act even if it the world is not indeterministic. A careful demon could influence individual molecular motions in such a way as to preserve the macroscopic properties of the system and so remain unobservable. Since the molecular storm drives most cellular processes, our demon could influence bodies and brains.

---

[42] Baudelaire (1998). Baudelaire describes the devil as being all in favour of the retreat of superstition. The only time he worried about his power was when he heard a preacher say, "La plus belle des ruses du diable est de vous persuader qu'il n'existe pas". The translation I'm using comes from the film Singer (1995).

Despite these possibilities, the life of a scientific demon would be quite dull. It might be able to control the behaviour of cells, hence that of creatures. It might manage a miracle, turning the sky dark thanks to a carefully organised storm. But making a ghost appear would be trickier.

Mischievous demons. A mischievous demon does exactly what it wants, joyfully contradicting physical law whenever it feels like it, but in a way that can't be picked up by scientific methods. It may not be a coincidence that ghosts are only observed when people are alone and unequipped with recording devices. When an attempt is made to scientifically investigate their existence they retreat, laughing at human illusions of knowledge. More generally, to sidestep science, a mischievous demon just needs to make sure their actions are not reproducible. If all demonic acts are one-offs, there's not much science can do.

Deceiving demons. Descartes imagined that "…there is some deceiver or other, supremely powerful and cunning, who is deliberately deceiving me all the time."[43] In the face of this deceiver, he realised that the only certain thing was the existence of the deceived, hence "I think therefore I am". He seems to have been so horrified by the implications of this that he quickly invokes a benign supernatural being to guarantee the veracity of his sensory impressions. But in the absence of such a guarantee, a demon is free to do what it likes with us. This is close to the brain in a vat thought experiment or the argument that we live in a simulation.

A more hands-off way for such a demon to operate would be to design our senses and intuitions from the start so that we are incapable of seeing its actions. We only know the world through our senses; control of them is control of everything. While we experience the universe as governed by laws, there may be demons all around us, but we are wired so that they and their actions are invisible to us. This might be termed paranoid radical scepticism.

13.16   The Limits of Physics 4

Austere physicalism is constrained by the general limits on physics. Chapter 4 showed how representational understanding lets us escape the limits of imaginative understanding. An important part of this representational understanding is the ability to describe and model systems and Chap. 5 investigated the theoretical and the practical limits of simulation. The Church-Turing-Deutsch principle states that there are no theoretical limits, every physical

---

[43] Descartes (1998), second mediation.

system can be simulated. The only limit is our ability to build universal quantum computers.

At the start of Chap. 4, I took a sly dig at philosophers who bat around metaphysical truths which seem obvious until someone comes up with a convincing counter-argument. But the same is true of similar statements by scientists. It's fine to talk about the Church-Turing-Deutsch principle and the universality of reason it implies. But it is a conjecture vulnerable to counter-example.

Current physics is no more than the best model we have available and is unlikely to be a final theory. Most physicists would accept that the physics we currently think of as fundamental is an effective theory, an approximation valid in the range of energies we can explore.

But physics may be an effective theory a much deeper sense. We've no way of knowing if all we observe and the rules that govern it are a low-dimensional projection of some vastly more complex space. The Church-Turing-Deutsch principle, if it applies at all, may apply just to our tiny corner of this space. We've no way of knowing whether different minds would have a different fundamental physics. Science helps us understand the limits of our imaginative understanding. We've no way of knowing if our representational understanding is similarly constrained. Fundamental physics itself may be another artefact of our cognitive limitations, another illusion.

We've no way of knowing if we are radically wrong about the nature of the world. John Heil describes Fodor asking us to imagine.

> … God creating the world. God calls together all his [sic] smartest angels. To one he assigns the task of working out laws of meteorology, to another the job of devising laws of geology, a third is dispatched to cook up laws of psychology, and so for every domain of the special sciences. To the smartest angel God assigns the task of working out the laws of basic physics. 'But', God enjoins, 'don't get in the way of those other angels!'[44]

We've no way of knowing whether we might be just one more random string produced by a mindless Great Programmer. And that's that. It's not possible to go any further. Austere physicalists need also to be metaphysically austere.

---

[44] Heil (2003).

13.17   Further Reading

For a rousing defence of eliminativism, see Rosenberg (2022). A radical statement of the eliminativist position from a philosophical perspective is in Horgan and Potrč (2008); Ladyman and Ross (2009) presents a similarly sparse metaphysics. Churchland (2013) is a book-length overview of eliminativism and mental events. For a discussion of eliminative materialism and the British Emergentists, see Ramsey (2022), sec. 1. Rosenberg (2006), Chap. 1 has a lovely discussion of the politics of reductionism. For quantum monism, see Schaffer (2010) or Calosi (2018). As a way of thinking about purpose and the intentional stance, Braitenberg (2004) is peerless.

More suggestions for reading can be found at www.TheMaterialWorld.net.

References

Al-Khalili J, Chen EK (2024) The Decoherent Arrow of Time and the Entanglement Past Hypothesis. https://doi.org/10.48550/arXiv.2405.03418

Baudelaire C (1998) Le spleen de Paris: petits poèmes en prose. Librairie générale française, Paris

Born M (1989) Physics in my generation. Springer, New York

Braitenberg V (2004) Vehicles: experiments in synthetic psychology. MIT Press, Cambridge, Mass

Broad CD (1925) The Mind and Its Place in Nature. Harcourt, Brace and Company, New York

Calosi C (2018) Quantum monism: an assessment. Philos Stud 175:3217–3236. https://doi.org/10.1007/s11098-017-1002-6

Chan D (2017) The AI That Has Nothing to Learn From Humans. In: The Atlantic. https://www.theatlantic.com/technology/archive/2017/10/alphago-zero-the-ai-that-taught-itself-go/543450/. Accessed 9 Jun 2024

Chen B, Huang K, Raghupathi S, et al (2022) Automated discovery of fundamental variables hidden in experimental data. Nat Comput Sci 2:433–442. https://doi.org/10.1038/s43588-022-00281-6

Churchland PM (1981) Eliminative Materialism and the Propositional Attitudes. The Journal of Philosophy 78:67. https://doi.org/10.2307/2025900

Churchland PM (2013) Matter and consciousness. MIT Press, Cambridge, Massachusetts

Cornman JW (1968) On the Elimination of "Sensations" and Sensations. The Review of Metaphysics 22:15–35. https://philpapers.org/rec/COROTE

Dennett DC (2013) Intuition pumps and other tools for thinking. W.W. Norton & Company, New York

Dennett DC (2017) From bacteria to Bach and back: the evolution of minds. W.W. Norton & Company, New York

Descartes R (1998) Discourse on method ; and: Meditations on first philosophy. Hackett Pub, Indianapolis

Drake S (1960) The Controversy on the Comets of 1618: Galileo Galilei, Horatio Grassi, Mario Guiducci, Johann Kepler. University of Pennsylvania Press

Edis T, Boudry M (2014) Beyond Physics? On the Prospects of Finding a Meaningful Oracle. Found Sci 19:403–422. https://doi.org/10.1007/s10699-014-9349-z

Ellis G (2016) How can physics underlie the mind? Springer Berlin Heidelberg, New York, NY

Gillett C (2018) Reduction and emergence in science and philosophy. Cambridge University Press, Cambridge

Graham L (2023) Molecular storms: the physics of stars, cells and the origin of life. Springer, Cham, Switzerland. https://doi.org/10.1007/978-3-031-38681-7

Heil J (2003) Levels of Reality. Ratio 16:205–221. https://doi.org/10.1111/1467-9329.00218

Horgan T, Potrč M (2008) Austere realism: contextual semantics meets minimal ontology. The MIT Press, Cambridge (Mass.)

Kim J (2007) Physicalism, or Something Near Enough, Princeton University Press, Princeton (NJ)

Kitcher P (1995) The advancement of science: science without legend, objectivity without illusions. Oxford University Press, New York

Ladyman J, Ross D (2009) Every thing must go: metaphysics naturalized. Oxford University Press, Oxford

Laplace PS (1995) A philosophical essay on probabilities. Dover Publications, New York

Laughlin RB (2005) A different universe: reinventing physics from the bottom down. Basic Books, New York

Liu C (2014) The three-body problem. Tor Books, New York

Longino H (2019) The Social Dimensions of Scientific Knowledge. The Stanford Encyclopedia of Philosophy. https://plato.stanford.edu/entries/scientific-knowledge-social/

Melnyk A (2020) Physicalism. In: Pritchard D (ed) Oxford Bibliographies in Philosophy. Oxford University Press, New York

Merricks T (2006) Objects and persons. Clarendon Press, Oxford

Ramsey W (2022) Eliminative Materialism. The Stanford Encyclopedia of Philosophy. https://plato.stanford.edu/entries/materialism-eliminative/

Rorty R (1965) Mind-Body Identity, Privacy, and Categories. The Review of Metaphysics 19:24–54. https://philpapers.org/rec/RORMIP

Rosenberg A (1991) How is Eliminative Materialism Possible. In: Bogdan RJ (ed) Mind and common sense: philosophical essays on commonsense psychology. Cambridge University Press, Cambridge; New York

Rosenberg A (2006) Darwinian reductionism, or, how to stop worrying and love molecular biology. University of Chicago Press, Chicago

Rosenberg A (2022) How to be an Eliminativist. Philosophical Aspects of Origin 19,1. https://doi.org/10.53763/fag.2022.19.1.198

Schaffer J (2010) Monism: The Priority of the Whole. The Philosophical Review 119:31–76. https://doi.org/10.1215/00318108-2009-025

Searle JR (2007) Freedom and neurobiology: reflections on free will, language, and political power. Columbia University Press, New York

Sellars W (1992) Science, perception and reality. Ridgeview Publ, Atascadero, Calif

Singer B (1995) The Usual Suspects, Polygram Filmed Entertainment

Tahko TE (2021) Unity of science. Cambridge University Press, Cambridge

Thomasson AL (2010) Ordinary objects. Oxford University Press, Oxford New York

Van Inwagen P (2014) Existence: essays in ontology. Cambridge University Press, Cambridge

Van Inwagen P (1995) Material beings. Cornell University Press, Ithaca, NY

Weinberg S (1992) Dreams Of A Final Theory: The Search for The Fundamental Laws of Nature. Vintage Books, New York

Wilson JM (2021) Metaphysical Emergence. Oxford University Press, Oxford

14

Eliminating Emergence

Summary Our intuition tells us that "more is different". Yet we have seen that emergence is an illusion. How can these be reconciled? This chapter takes a physicalist approach to this question by unpacking the concept of emergence. Our perception of phenomena as emergent can be seen as a shorthand for one or more of the following: wonder at the range of interesting phenomena; lack of imaginative understanding; unsolved problems; the difficulty of prediction in complex systems; the properties of models and approximations or the nature of the special sciences.

We know that the term emergence is redundant, but what can we learn when it is used? What can we make of our intuition that "more is different"? This chapter uses austere physicalism to show that the concept of emergence is a shorthand for one or more of the following: sheer wonder at the ubiquity and range of interesting phenomena; a lack of imaginative understanding; unsolved problems; the difficulty of prediction in complex systems; the properties of models and approximations and the practice of science. The next chapter will apply this framework to the examples of Chap. 2.

Not only is the term emergence redundant, but using it tends to conceal interesting things. As an illustration, let's return to the example of different balls rolling down a slope. They can be described as MR emergent in the sense that pretty much whatever the balls are made of, they seem to behave in the same way. Eliminating the term allows us to see how the motion of every ball is different. Due to the interaction of the molecular structure of the ball's surface with that of the slope. Due to their not being perfectly homogenous

© The Author(s), under exclusive license to Springer Nature Switzerland AG 2025
L. Graham, *Physics Fixes All the Facts*, The Frontiers Collection,
https://doi.org/10.1007/978-3-031-69288-8_14

or perfectly spherical. It allows us to ponder how our senses are too limited to observe the relevant differences between the balls or the tiny differences in the nature of their motion. And we can refer to Newton's Laws that govern their motion as an effective theory, a low energy approximation to special relativity.

14.1   Interesting Physics at All Scales

One of the most astonishing things about the world in which we live is that there seems to be interesting physics at all scales. Whenever we look in a previously unexplored regime of distance, time, or energy, we find new physical phenomena. From the age of universe, about 10^{18} sec, to the lifetime of a W or Z, a few times 10^{-25} sec, in almost every regime we can identify physical phenomena worthy of study[1]

Sometimes the term emergence captures no more than the sense of wonder expressed in this passage. There is interesting physics at all scales because higher levels are often largely independent of the details of their makeup. This means we can observe approximate empirical regularities between macroscopic objects, so can use effective theories instead of needing to solve everything from the bottom up. Without this, it's hard to see how science could ever start.

Why is the world like this? As I discussed in Chap. 8, part of the explanation is that we are macroscopic creatures who make massively coarse grained observations. Beyond that, we are back to a fine tuning argument that without such a property complex systems could not exist. Which is no more than saying "That's just the way it is".

14.2   Bafflement

Modern physics does not make sense. The non-locality of quantum physics is the most obvious example. Yet this non-locality is no challenge to physicalism since it is physics. As we will see in the next chapter, entanglement also lies behind many of the most interesting phenomena in condensed matter physics.

---

[1] Georgi (1993).

The other part of modern physics, general relativity, seems less puzzling. We are used to being taught it by analogy with ball bearings and rubber sheets and one of its implications can be tested with a fairly ordinary telescope as Dyson and Eddington did in 1919.[2] Yet dig deeper and it is just as baffling as quantum physics. For example, it implies the whole history of the universe exists as an unchanging block in 4D space–time and there is no reason to privilege one direction of time over another.

We have as good representational understanding of all this as we do of anything. Yet however well we know the maths, we remain imaginatively closed to it. This is what some uses of the term emergence tell us. Then there are some more specific senses of imaginative closure which I shall turn to in the next section.

14.3 Imaginative Closure

Cognitive limitations are behind everything in this chapter. If you ask why we use particular models or why particular problems are unsolved, the final answer will be to do with the capacity of our minds. A demon would see the undifferentiated blobject. We have to slice it up to make it comprehensible. In this section, I want to concentrate on a few more specific instances in which our imaginative closure can clarify particular ways in which the term emergence is used.

The basic idea is neatly captured in this dialogue between an exponent of reductionism (PRO) and an opponent (CON):

> PRO: Why do you need your coarse-grained description, which is supposed to hold only on a single scale, the scale under consideration, if the supercomputer evolves the formation of larger and larger composed objects out of the fundamental ones? Let it go on and simulate humans.
>
> CON: I need it precisely for my understanding in terms of simple mechanisms. I need this kind of understanding to abstract universal features from different realizations, to be inspired to new ideas, and in particular to design the new computer generations you are looking forward to.[3]

We are coarse grained creatures who need coarse grained descriptions. Why a particular set of descriptions, or, if you like, a particular set of special

---

[2] Dyson et al. (1920).
[3] Meyer-Ortmanns (2014).

sciences? Stephen Blundell gives one answer with an interesting parallel between concepts of emergence and storytelling:

> …emergent narratives [capture] the essence of reality in a way that is far better fitted to the constraints and preferences of the human mind than a brute description of all the details at the lowest level.[4]

Emergence then becomes an efficient way of describing elements of reality to fit within the limits of our imaginative understanding.[5] Blundell asserts that his narratives are "elements of reality". I find this peculiar. Of course, any narrative that exists in a mind is a real property of humans but this applies to supernatural beings too. For believers, they play an important role in making the world comprehensible. For an observer, they are necessary to understand the behaviour of believers. But they are properties of minds not of the world. To say any different is to fall into the mind projection fallacy.

In a 2010 paper,[6] Brian Johnson takes a different approach by focussing on a specific cognitive constraint. He argues that we are unable to think consciously in parallel and demonstrates this using a simple example. Try to simultaneously read and understand the first and last sentences of this paragraph. You can't do it; no-one can do it. The brain can do many things in parallel but this isn't one of them. Emergent systems are characterised by multiple simultaneous interactions which, without the capacity to think in parallel, are impossible to follow. It only takes a few elements to overwhelm our cognitive ability so we find them surprising and mystifying.

This is a useful unpacking of the idea of imaginative understanding. We simply are not built to be able to "get our heads round" complex systems. As with the Game of Life, while we may have perfect representational understanding we will always need simplified explanations to make sense of it. Our imaginative understanding can be helped using cognitive scaffolds. My simulations of the Ising model in Sect. 10.4 were designed to do exactly that, making it easier to understand phase transitions. But the nature of our minds means there will always be a gulf between imaginative and representational understanding.

---

[4] Blundell (2017).

[5] Rosenberg (2011), Chap. 1 contains an interesting discussion of the importance of narratives for imaginative understanding.

[6] Johnson (2010).

14.4 Unsolved Problems

Elanor Taylor defines an emergent phenomenon as one involving unsolved problems.[7] But the idea is implicit in many of the other definitions: things are called emergent until we learn the physics.

Everything seems like magic until you know how it's done. Think of a rainbow. This is obviously an emergent property of sunlight and rain, completely different from its two constituents. Planetary motion is obviously the result of the divinity of heavenly bodies and the perfection of spherical motion. Endless more examples can be found throughout the history of science.

It's worth unpacking a bit further this sense of emergence. We know that everything from molecules to tables is made of quantum fields. We think we know the maths that describes them since standard quantum physics works fine as long as we are not interested in extremely high energies or extremely strong gravitational fields. But for the moment, quantum many-body calculations are computationally intractable for more than a few dozen particles.

Whether this will remain the case is anyone's guess. One avenue for optimism is the rapid development of quantum computing bringing with it the possibility of quantum simulators. Another is the possibility of finding entirely new methods of computation. More prosaically, but also more importantly, computational methods are constantly improving. Remember the discussion of the travelling salesperson problem in Box 5.2? We saw it is computationally intractable for more than 30 or so nodes. But it has been solved for 50,000 nodes using clever tricks. There's no reason that such workarounds cannot be found for problems in quantum physics. Indeed, this is a large part of the discipline of quantum chemistry. The scope of such bottom-up approaches are also limited by the issue of contingency and predictability discussed in the next section.

There are unsolved problems at every level of science. Wikipedia lists over 100 significant unsolved problems in physics alone. As an example, take non-equilibrium statistical mechanics which describes the irreversible processes that constitute almost everything that happens everywhere. Since Prigogine's Nobel prize winning work on near-equilibrium thermodynamics in the 1970s, there has been constant progress. One of my favourite examples is Jeremy England's work on dissipative adaptation.[8] But we are a long

---

[7] Taylor (2015).
[8] England (2013).

way from anything which looks like a general theory of non-equilibrium behaviour.

Will such problems remain forever unsolved? Using the terminology introduced in Chap. 4, this is the same as asking if we are representationally closed to some things. These are questions without answers. The Church-Turing-Deutsch principle (Sect. 5.7) implies that everything can be simulated. But even if it were proved true, I don't think it implies that there are no limits to representational understanding. Anyone who has built an agent based model or trained an AI will know that there is a big difference between creating a model and understanding its outputs.

My guess is that even if we find we are representationally closed to some things, we will build machines without such limits. When they learn about the New Mysterianism and the "hard" problem of consciousness, it's easy to imagine them chuckling among themselves at the way their creators still saw themselves as the centre of the universe. We may even be able to persuade them to explain to us what they learn in simple terms, using their knowledge to build minimal models tailored to the limits of our understanding.

Box 14.1 Misunderstanding the world

In a 2020 paper,[9] Adrian Kent shows how a limited understanding of microscopic rules can lead us to profoundly misunderstand the macroscopic nature of the world. The paper adds a probabilistic element to the Game of Life, randomly flipping squares and then studies the resulting dynamics of gliders. If the flip happens a long way from a glider, it will most likely have no effect on the glider's path. The closer the flip is to a glider, the more likely it is to have an effect. If it happens close to a group of interacting gliders, it may affect all of them.

He then adds a further rule which makes these flips to some extent dependent on the later states of the gliders in a way that stabilises them: making flips close to the gliders a bit less likely than flips far away. He argues that to an observer who does not know these rules, the model will show both downward causation (the pattern of the gliders affecting the micro-distribution of flips) but also violate causality with the future seeming to affect the past (since the probability of errors in early periods depends on the pattern of the gliders later in time). Kent takes this to have implications for the way consciousness could have real effects. I take it to show how emergent phenomena are illusions caused by our cognitive limitations.

---

[9] Kent (2020).

14.5 Explaining and Predicting

Understanding a system and being able to predict its evolution are two different things. We've already seen an example in the three-body gravitational system of Sect. 7.1. Since the equations are the same as apply to Newton's apple, it would be hard to argue that the system is not fully understood. But in the general case, it is impossible to predict.[10]

This is a characteristic of chaotic systems. Without the ability to measure the initial state of a system to infinite precision, its behaviour becomes less and less predictable with time. There may be regions of stability, called attractors, where behaviour is stable, but the system can suddenly jump from one such state to another, or into an unstable state.

Possibly the simplest system in which chaotic behaviour can be seen is the dyadic map. This is an algorithm for producing a series of numbers between 0 and 1:

1. Start with a number between 0 and 1
2. Double it
3. If the result is between 0 and 1, go back to step (2)
4. If the result is greater than or equal to 1, subtract 1 then go back to step (2)

The algorithm has the fascinating property that if the initial number is rational (i.e. can be written as the ratio of two integers), the series will eventually settle down either to zero or to a repeating pattern. If the initial number is irrational, the series goes on for ever without repeating. To see this, note that any rational number has a binary representation that either has a finite number of digits or that repeats a patten of digits for ever. The map progressively chops off digits in the binary representation. For example, in binary 3/4 is 0.11. Multiplying by 2 shifts the digits one place to the left, giving 1.1, then subtracting 1 gives 0.1. Multiplying by 2 gives 1, subtracting 1 gives 0. For a rational number, this process will either end in zero or a pattern that repeats for ever. An irrational number has no finite representation so the process will generate a sequence that goes on forever without repeating.

Since almost all numbers between 0 and 1 (or in any interval) are irrational, there is always an irrational number arbitrarily close to every rational number. This means that an arbitrarily small change in the initial value can lead to the result shifting from a series that settles down to one that never

---

[10] Lack of predictability features in some definitions of emergence, notably inferential emergence, see Humphreys (2008).

repeats. The series of numbers produced by the algorithm are exquisitely sensitive to the starting value. This is the hallmark of a chaotic system.

Chaotic behaviour occurs in deterministic systems. This means they are predictable in principle. Unpredictability in practice arises from our inability to precisely measure initial conditions. Think of a pinball machine. It is fully described by Newtonian dynamics and whatever deterministic rules control the behaviour of its elements. Yet play for long enough and the position of the ball will become effectively random (or at least random conditional on the rules of the game).

Chaos is an evocative term. Yet chaotic systems are not entirely unpredictable. There are a host of statistical techniques which can help understand them. This is neatly summarised in the title of a 2006 book on the subject: "Bigger than chaos: understanding complexity through probability".[11] And there is continuous progress. In terms of weather forecasting, despite butterflies flapping their wings all over the world, the UK Meteorological Office claims that "our four-day forecast is now as accurate as our one-day forecast was 30 years ago".[12] What's more, chaotic systems may be ripe for the applications of AI techniques. A 2023 paper finds they allow much longer-range forecasts than classical methods.[13]

The Mandelbrot set is the canonical example of a complex system in which the outcome depends exquisitely on the initial conditions. John Hubbard, in a lecture from 1987, turns the complexity round, saying that rather than being beyond our understanding:

> It is therefore a real message of hope, that possibly biology can really be understood in the same way that these pictures can be understood.[14]

He goes on to make the parallel between the simple rules that lead to the set and the relatively simple program in DNA that leads to humans.

Understanding a system and being able to predict it are two different things. Emergence blurs this distinction. Does unpredictability mean that some things are beyond physics? If so, we would be forced to conclude the dyadic map is beyond arithmetic.

---

[11] Strevens (2003).

[12] https://www.metoffice.gov.uk/about-us/who-we-are/accuracy.

[13] Gilpin (2023).

[14] A video of the lecture can be found at www.TheMaterialWorld.net.

14.6 Models and Approximations

In Chap. 10, I showed that emergence is often not about physical systems but about the models and approximations scientists use to describe them. Studying such models is an interesting exercise which teaches us about the practice of science, the interests of scientists and the cognitive and computational resources that they have available. I also discussed how apparently discontinuous emergent properties are continuous if we look closely enough.

These are theoretical and empirical sides of the same coin. Emergence used in this sense confuses the nature of science and our cognitive structure with the nature of reality.

14.7 The Special Science Stance

A further way in which the term emergence confuses the practice of science with the nature of reality is about the diversity of the special sciences. I have already quoted this passage from Bishop:

> Our argument will be that the plethora of explanatory pluralism in the sciences turns out to be good evidence for contextual emergence.[15]

But I've argued that the structure of the special sciences is arbitrary and constantly changing, just our latest way of slicing up the blobject as a function of our interests and our resources. In other words, it is a stance, a way of simplifying reality to make it more comprehensible to us.

Of course, higher level descriptions are often explanatorily more powerful than lower level ones. Daniel Dennett argued in a 1991 paper that macro-objects are *real patterns* whose reality depends on their explanatory power and predictive ability. The most obvious example comes from the Game of Life: talking of a glider is much more useful than giving a list of coordinates of its cells and leaving you to do the calculations in your head. Dennett gives a more evocative example:

> Predicting that someone will duck if you throw a brick at him is easy from the folk-psychological stance; it is and will always be intractable if you have to trace the photons from brick to eyeball, the neurotransmitters from optic nerve to motor nerve, and so forth.[16]

---

[15] Bishop et al. (2022), p. 23.
[16] Dennett (1991).

The assertion that such patterns are real is just as peculiar as in the case of Blundell's narratives that I discussed in Sect. 14.3. The intentional stance is only real as a property of a human brain.

Using utility as a criterion for reality has the further off-putting consequence of meaning reality is constantly changing. We can read the structure of reality from neither the models nor the language used by scientists. To attempt to do so is to fall into the mind projection fallacy.

14.8 Discussion

After all that, the only use of the term emergence that remains meaningful is as a synonym for non-fundamental. We've already encountered this in the distinction between the fundamental mass of nucleons, given by the Higgs field, and the emergent mass that results from quantum chromodynamic interactions. Another example would be suggestions that spacetime is emergent, arising from interactions at a more fundamental level. But such usage is restricted to the highest energy physics. Outside of this context, the term can be discarded without any loss of meaning. Still, reflecting on why a phenomenon was described as emergent can be interesting and the concepts explored in this chapter will help clarify what is really going on.

14.9 Further Reading

For reviews of the literature on scientific progress, see Rowbottom (2023), Chap. 1 or the Stanford Encyclopaedia of Philosophy, Niiniluoto (2024). For a wonderful popular introduction to chaos, see Gleick (1987). An interesting discussion of the link between predictability and computational emergence is in Tabatabaei Ghomi (2022).

References

Bishop RC, Silberstein M, Pexton M (2022) Emergence in context: a treatise in twenty-first century natural philosophy. Oxford University Press, Oxford, United Kingdom

Blundell SJ (2017) Emergence, causation and storytelling: condensed matter physics and the limitations of the human mind. Philosophica 92. https://doi.org/10.21825/philosophica.82114

Dennett DC (1991) Real Patterns. The Journal of Philosophy 88:27. https://doi.org/10.2307/2027085

Dyson FW, Eddington AS, Davidson C (1920) A determination of the deflection of light by the sun's gravitational field, from observations made at the total eclipse of May 29, 1919. Phil Trans R Soc Lond A 220:291–333.https://doi.org/10.1098/rsta.1920.0009

England JL (2013) Statistical physics of self-replication. The Journal of Chemical Physics 139:121923. https://doi.org/10.1063/1.4818538

Georgi H (1993) Effective Field Theory. Annu Rev Nucl Part Sci 43:209–252. https://doi.org/10.1146/annurev.ns.43.120193.001233

Gilpin W (2023) Model scale versus domain knowledge in statistical forecasting of chaotic systems. Phys Rev Res 5:043252. https://doi.org/10.1103/PhysRevResearch.5.043252

Gleick J (1987) Chaos: making a new science. Viking, New York

Humphreys P (2008) Computational and Conceptual Emergence. Philos of Sci 75:584–594. https://doi.org/10.1086/596776

Johnson BR (2010) Eliminating the mystery from the concept of emergence. Biol Philos 25:843–849.https://doi.org/10.1007/s10539-010-9230-6

Kent A (2020) Toy Models of Top Down Causation. Entropy 22:1224. https://doi.org/10.3390/e22111224

Meyer-Ortmanns H (2014) On the Success and Limitations of Reductionism in Physics. In: Why more is different. Springer, New York

Niiniluoto I (2024) Scientific Progress. The Stanford Encyclopedia of Philosophy. https://plato.stanford.edu/entries/scientific-progress/

Rosenberg A (2011) The atheist's guide to reality: enjoying life without illusions. W.W. Norton, New York

Rowbottom DP (2023) Scientific Progress. Cambridge University Press

Strevens M (2003) Bigger than Chaos: Understanding Complexity through Probability. Harvard University Press, Cambridge, Mass

Tabatabaei Ghomi H (2022) Setting the Demons Loose: Computational Irreducibility Does Not Guarantee Unpredictability or Emergence. Philos sci 89:761–783. https://doi.org/10.1017/psa.2022.5

Taylor E (2015) An explication of emergence. Philos Stud 172:653–669. https://doi.org/10.1007/s11098-014-0324-x

15

More is Fascinating

Summary This chapter returns to the examples of Chap. 2 and shows how they can be understood in a physicalist framework without resorting to the concept of emergence. It ends with a challenge. If you have a system which you claim is emergent, there is a simple procedure you can follow to substantiate your claim and convince an austere physicalist that you are right.

Let's return to the examples of Chap. 2. I'm going to work through them one by one using the framework of the previous chapter to show how they can be thought about without using the concept of emergence. Often, this will involve discussing simulations and I have argued at various points that simulations are valid reductionist explanations.

When you reach the end of each section, I invite you to ask yourself: what would using the term emergence add to what you've read? What have I missed out that would be captured by the term? A different objection would be that this chapter consists of words and not quantum field theory, so fails to fully deliver on the promise of physicalism. I'll turn to this in the penultimate section.

And if you have an emergent system in mind which you think my framework fails to fit, the chapter ends by giving a straightforward procedure which you can use to convince an austere physicalist of your case.

© The Author(s), under exclusive license to Springer Nature
Switzerland AG 2025
L. Graham, *Physics Fixes All the Facts*, The Frontiers Collection,
https://doi.org/10.1007/978-3-031-69288-8_15

15.1 Protons and Neutrons

Quantum chromodynamics (QCD) describes the strong interaction between quarks. While there are still unsolved problems, it has so far successfully passed all experimental tests. Its short-range nature and the fact that gluons themselves have color means it is highly non-linear and standard perturbation-based calculation methods cannot be applied. However simulation techniques are extremely successful. A method called Lattice QCD can solve the theory to a level of precision limited only by the available computing power.[1] One of its big successes has been to theoretically determine the mass of a proton to within a few percent.[2] There seems little doubt that as computational methods and computing power improve, so will the accuracy of these simulations.

Any of the definitions of weak emergence would imply that nucleons are emergent from their component quarks. Yet QCD explains why at low energies, we can just deal with nucleons without worrying about their internal complexity. You could not hope for a better example of the remarkable power of physics.

15.2 The Classical World

The classical world emerges from the quantum world due to environmental decoherence. Quantum entanglement is the key to this process. When a particle from the environment scatters off the system in superposition, they become entangled and share a common wave function which is less coherent. If this happens repeatedly, coherence is lost.

This might seem like a canonical case of contextual emergence and indeed is treated as such by Bishop et al.[3] But it is a direct consequence of quantum non-locality. As such, it is standard physics. You may find such non-locality counterintuitive. You won't be the only one. At the end of a paper attempting to show something was missing from quantum physics, Einstein famously wrote:

No reasonable definition of reality could be expected to permit this[4]

---

[1] This technique is behind the visualisation at https://arts.mit.edu/projects/visualizing-the-proton/.

[2] Dürr et al. (2008).

[3] Bishop et al. (2022), Sec. 4.7.

[4] Einstein et al. (1935).

So much the worse for our intuition about what is reasonable.

15.3 Atoms and Molecules

In Chap. 2, I gave nuclei, atoms and molecules as three examples of "more is different". The equations of quantum physics cannot be solved exactly for even the simplest atoms. In general, only numerical solutions are available though only computing power limits their precision.

Turning now to molecules, we enter the realm of quantum chemistry. I've already discussed one example, the calculation of the bond angle of water (Sect. 11.3). Others are to be found from studies of the properties of enzymes and catalysts to the absorption of molecules at surfaces, from drug design to understanding photosynthesis. There is interesting work on how to apply quantum computing to quantum chemistry.

For all but the smallest molecules, simplifications and approximations are necessary. One common technique is known as the Born–Oppenheimer approximation. Nuclei are heavy, so move much more slowly than far lighter electrons. The approximation involves solving for the wave functions of electrons while keeping the position of the nuclei fixed. Repeating this for different nuclear positions gives an approximation to the potential in which the nuclei move. Since the computational complexity of the problem scales with a power of the number of particles, splitting the problem into smaller ones is computationally efficient.

The proponents of contextual emergence refer to the Born–Oppenheimer approximation as a "stability condition",[5] apparently basing their claim on earlier work by the Swiss chemist Hans Primas.[6] But this is yet another example of confusing the map and the territory. The Born–Oppenheimer approximation is a feature of a particular approach to modelling. It is not a property of the world. Nature needs no approximations.

Quantum chemistry is a triumph of quantum physics and computational techniques. In 1927, Walter Heitler and Fritz London published the first quantum mechanical description of the hydrogen model.[7] Less than 90 years later, a 2014 paper[8] calculated a range of properties for 134,000 small molecules made up of the elements carbon, oxygen, hydrogen, nitrogen and

---

[5] Bishop et al. (2022), sec. 4.6.2 and Bishop (2019), sec. 4.4.3.
[6] Primas (1998).
[7] Heitler and London (1927).
[8] Ramakrishnan et al. (2014).

fluorine, claiming results at or better than experimental accuracy. If you want to find examples of non-reductive physicalism, you need to look elsewhere.

15.4 Chemical Oscillators

Chemical oscillators can produce rich sets of complicated patterns. As such, they seem a natural candidate for emergence. Section 2.4 gave the example of the Belousov-Zhabotinsky or BZ reaction. The reaction involves 30 or so different chemicals and is poorly understood. Yet we can get an idea how it works in terms of the autocatalytic reactions discussed in Sect. 9.5. Imagine we have an autocatalytic reaction which transforms a pink reactant into a blue product which is also a catalyst. Then another reaction transforms the blue chemical into a colourless end product. We also need to assume that the pink chemical is slow to diffuse through the dish.

1. When we initially prepare the reactants, the dish is pink.
2. A random fluctuation produces some of the blue product.
3. This then catalyses the first reaction and sets of the exponential autocatalytic process.
4. This rapidly uses up the pink chemical producing a blue spot.
5. Then the second reaction gradually turns the blue spot transparent.
6. At the same time, the pink chemical diffuses back in from the surroundings allowing the process to start again.

Does this sound familiar? It is an example of predator–prey dynamics. Exactly the same description can be used to explain the population dynamics of, say, lions and gazelles. As gazelles reproduce, there is more food for lions so their population increases too. A higher population of lions will tend to lower the population of gazelles. Given time-lags due to reproduction, both populations show regular cycles.[9]

In Chap. 2, I cited Prigogine: "To change color all at once, molecules must have a way to 'communicate.' The system has to act as a whole".[10] But molecules communicate no more than do lions and gazelles. The holistic dimension of the reaction is an illusion.

---

[9] Chemical oscillators and interactions between lions and gazelle are described by the Lotka-Volterra equations, a good example of a minimal model that applies to many different systems. Of course, like any model, it only applies approximately.

[10] Prigogine and Stengers (1984), p. 148.

15.5 Symmetry Breaking

Remember the finely-balanced ball that could roll left or right into two identical valleys? This is the simplest example of the distinction between explanation and predictability of Sect. 14.5. Symmetry breaking is one place where contingency enters physics. We may be able to work out, at least in principle, why the ball went the way it did, or we may not. But in either case, after a long sequence of such symmetry breakings it may be impossible to keep track of them. Despite our understanding the physics perfectly, we may not be able to explain why the ball is here, rather than there.

Then came phase transitions and their universal aspects as examples of symmetry breaking. I discussed these in the context of the thermodynamic limit (Sect. 10.1). The key point is that in finite systems phase transitions are not discontinuous but gradual and universality is only approximate. Because the systems we study typically contain large number of molecules, they are good approximations. But they remain approximations. This is true generally. Spontaneous symmetry breaking is important throughout physics, but it too only occurs in infinite systems.

I cannot improve on Kadanoff's characterisation: infinitely more is different; more is the same.[11]

15.6 Quasiparticles

In condensed matter physics, the complex interactions between large numbers of atoms can sometimes be represented as quasiparticles. Phonons represent particular vibrational modes of a crystal lattice. A single phonon captures the behaviour of the hugely complex underlying interactions.

The RGT shows how the universal properties of phase transitions depend only on some general properties. Phonons are often described in similar terms, with their properties depending only on the general symmetry properties of the underlying material. However, just as with the RGT this assumes the thermodynamic limit. Remove this assumption, and the properties of phonons will depend on the size of the underlying material, the nature of its boundaries and its precise constitution. We might be able to ignore this dependence to a high degree of approximation, but this is just a result of dealing with systems large enough to make the resulting errors small.

---

[11] Kadanoff (2009).

Often the whole point of studying phonons is to focus on how their properties relate to those of the underlying material. And specifically to design material so as to tune their phonons to achieve certain properties in terms of heat or sound conduction. Such exercises would be meaningless if phonons were independent of everything but general symmetry properties. This is from a 2022 study of phonons in nanoscale systems:

> This spatial confinement resulting from the reduction of bulk material down to 'finite' sizes affects a wide array of physical properties such as the phonon density of states, group velocity, specific heat capacity, and electron–phonon and phonon–phonon interactions, among others.[12]

As a parallel, take a guitar string. Whatever it is made of, within broad limits, you can tune it to give a particular note. But steel strings sound different from gut strings. While the note is the same, the tone depends on the material. If you use analyse the tone, you'll find that every string gives a different combination of frequencies. In principle, if you change a single atom in the strong, you'd be able to spot the difference in the frequency decomposition. Though in practice, such small differences would be swamped by thermodynamic noise. The same is true for phonons. Whether this matters depends on the questions we are asking.

In 1988, Michael Fisher wrote an article with the provocative title "Condensed Matter Physics: Does Quantum Mechanics Matter?". He writes in the conclusion that the aim of the paper is to give

> …a picture of the multifaceted character of modern condensed matter physics and the varying degrees to which quantum mechanics is of direct relevance or **almost** total irrelevance.[13]

Does quantum physics matter? For the properties of phonons in macroscopic materials, it is almost irrelevant in the same way the position of an atom is irrelevant to the behaviour of a guitar string. However it is directly relevant for phonons at the nanoscale.[14] I speculate that this is also true for other quasiparticles.

Superconductivity is a slightly different example since Cooper pairs are not usually described as quasiparticles. The theory that describes it is known as BCS after its creators Leon Cooper, John Bardeen and John Schrieffer.

---

[12] Ng et al. (2022).
[13] Fisher (2016). Emphasis added.
[14] For a review, see Chen (2021).

Like any good model, it involves some dramatic approximations. Here's Cliff Burgess:

> ...the BCS theory of superconductivity ... which ignores most of the mutual interactions of electrons, focussing instead on a particular pairing interaction due to phonon exchange. Radical though this approximation might appear to be, the theory works rather well (in fact, surprisingly well), with its predictions often agreeing with experiment to within several percent.[15]

This "several percent" is roughly the same accuracy to which universal properties of phase transitions are known. It is extraordinary how much simple models, like the RGT for universality or BCS for superconductivity, can tell us about the world. But they remain models.

This starts to make quasiparticles sound like useful fictions rather than elements of reality. This is Ellis's answer to whether quasiparticles are real:

> From the perspective of quantum field theory, the answer is a resounding yes. Each of these particles emerges from a wave-like description in a manner that is entirely analogous to that of photons. These emergent particles behave like particles: you can scatter other particles of them. ... Yes, they are the result of a collective excitation of an underlying substrate. But so are 'ordinary' electrons and photons, which are excitations of quantum field modes.[16]

There is a problem with this. The phonon description and the many-body descriptions of a lattice are equally valid. It just happens that, for our purposes, one is simpler than the other. You can use whichever is more appropriate to the problem you are solving without having to know anything about the other.

However, this isn't the case for wave/particle duality in quantum physics. If you want to explain the double-slit experiment for example, you need the wave description to explain the interference and the particle explanation to explain the points on the screen. Does this generalise to quantum field theory? As I mentioned back in Sect. 3.2, the status of particles in QFT is quite fragile. But they seem necessary to explain observations. Brigitte Falkenburg writes:

---

[15] Burgess (2004).
[16] Ellis (2020).

One follows Max Born, claiming pragmatically that in any quantum process waves propagate (and should therefore be prepared in approximately pure momentum states, in experiments), whereas particles are detected[17]

This weakens Ellis's analogy between quasiparticles and real particles. But whatever the answer is, it doesn't change much for physicalism. Either quarks and electrons are real and quasiparticles are just helpful approximations. Or the fundamental level consists of fields and everything else is just an approximation. I'll take either one.

15.7   The Quantum Hall Effect

A few sections back I discussed decoherence as an explanation of the separation between the quantum microscopic world and the classical macroscopic world. Many of the interesting phenomena in condensed matter physics transcend this distinction. Superconductivity and the quantum Hall effects are examples of *macroscopic* quantum systems. The combination of ultra-low temperatures and reduced dimensionality allows quantum behaviour to be relevant at everyday scales.

Theoretical explanations of both integer and fractional quantum Hall effects came soon after their discovery. While there are many open questions, I think it would be reasonable to say that the integer effect can be explained in a model of independent electrons whereas the fractional effect requires correlations between many electrons. Laughlin's Nobel Prize-winning model of the fractional effect involved writing down a single wave function for all the electrons in the sample. In general, topological effects are modelled as involving long-range entanglement between the particles in the system.[18]

In Chap. 11, I argued that while entanglement fitted the definition of strong emergence it was no challenge for physicalism since it is the heart of modern physics. Exactly the same is true of these models of topological order. Yes, it is impossible, in fact it is meaningless, to talk about the properties of one of the particles in the system without talking about the system as a whole.

If you accept that EPR experiments showing entanglement are fully described by quantum physics, then so too are these condensed matter

[17] Falkenburg (2007), p. 332.
[18] Chen et al. (2010).

phenomena.[19] They cannot be a challenge to a physicalist account. Entanglement seems magical. Long-range entanglement in condensed matter systems seem doubly magical. This represents yet another failure of our imaginative understanding.

The Hubbard model is a workhorse of condensed matter physics. Like the Ising model, it takes the form of a lattice, but allows sites to be either full or empty and particles to tunnel between different sites. In a fascinating paper[20] from 2021, Ioannis Kleftogiannis and Ilias Amanatidis present an elegant minimal model of the fractional quantum Hall effect in a 1D Hubbard chain. Strikingly, they show that even in systems with just a few particles, plateaux at fractional levels can arise. The simplicity of the model allows a detailed investigation of the mechanism of fractionalisation and the way in which topological effects depend on the interaction between the particles and the level of noise.

In Chap. 10, I showed how the Ising model clarifies the mechanism behind ordinary phase transitions. It seems to me that work like this has the potential to do the same for transitions between topological phases.

15.8 Bénard Convection

Modelling convection cells is a standard problem in fluid dynamics which can be found in many textbooks. However fluid dynamics is an effective theory par excellence. Its equations are high level, making no reference to molecular dynamics and only valid under specific conditions. Although derivations exist for restricted cases, there is as yet no general molecular level theory of the macroscopic behaviour of fluids.[21]

Despite this, simulation techniques have been extraordinarily successful in understanding fluids in general and convection cells in particular. A 1988 paper[22] showed how convection cells can arise in a system comprising of around 15,000 rigid discs. Nine years later, another paper[23] using Monte Carlo techniques extended this to around 20 million particles.

---

[19] At least, given our current understanding there is no clear evidence that they are not.

[20] Kleftogiannis and Amanatidis (2021).

[21] Bobylev (2018) is a discussion of the relation between the equations of fluid dynamics and Boltzmann's kinetic equation which underlies them.

[22] Rapaport (1988).

[23] Watanabe and Kaburaki (1997).

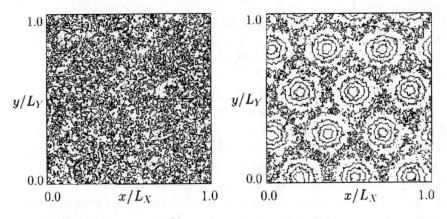

Fig. 15.1 Simulating convection[24]

Figure 15.1 is a graphical representation of the simulation's output. On the left is conduction, on the right convection. Compare this with my simulations of the Ising model in Sect. 10.4. In the same way, these simulations allow you to understand the particle-level details of how the phase transition, in this case from conduction to convection, arises. Like all simulations, they make approximations, but there seems little doubt that they allow a thorough understanding of fluid properties.

Bishop's sole authored book goes into Benard convection in some detail, concluding that.

> [it] is an instance of contextual emergence rather than reduction, where stability conditions provided by the large-scale fluid dynamics allow the existence of particular reference states and observables for convection[25]

A simulation allows you to see how these so-called stability conditions and large-scale dynamics arise from the interaction of molecular-level components.

---

[24] Reprinted Fig. 4 with permission from Watanabe and Kaburaki (1997). Copyright (1997) by the American Physical Society.
[25] Bishop (2019), p. 4.9.

15.9 Self-organisation

Around 10^{-32}s after the big bang, the universe contained a quark-gluon plasma. A short time later, this self-organised into protons and neutrons. Later still, these nucleons along with electrons self-organised into atoms. Even later, in clouds of gas, these atoms self-organised into molecules. If you want to take the story forward, return to my description of the early universe in Sect. 9.4 and insert "self-organised" where necessary. If you want to go still further, you could say how molecules on planetary surfaces self-organised to form more complex molecules, how these self-organised to form proto-cells. And so on to cells, evolution and all the way to Churchill's nose. For a physicalist, there is no other choice. Although it is beyond our imaginative understanding, the big bang sets into motion a set of physical processes which, 14 billion or so years later, leads the atom of copper being in Churchill's nose and these words describing it.

Self-organisation is everywhere. Like weak emergence, this means it is not a useful concept. And like weak emergence, it tells us not about the nature of the world but about our cognitive limitations. We don't usually refer to the first steps, involving quarks, atoms and molecules, as self-organisation. This is because we think we have imaginative understanding of them. In fact, I suspect this understanding is pre-quantum, based on Rutherford and Bohr's planetary model, but let's put that aside. The systems that are more typically described as self-organising are those for which we do not and cannot have imaginative understanding. Systems composed of large numbers of interacting agents are definitely not part of our manifest image. By now it should be no surprise that they continually enchant and baffle us.

If you want representational understanding, you need a simulation. As one example, in a 1987 paper,[26] Craig Reynolds developed an agent-based model of flocks in which individuals have just three rules: avoid collisions with nearby flockmates; attempt to match velocity with nearby; attempt to stay close to nearby flockmate.

Since Reynolds's work, the model has been refined and extended in many ways. A paper from 2009 compares a simulation of starling behaviour against data from actual flocks. While the paper makes quantitative comparisons, you can get an qualitative view from Fig. 15.2 which compares actual flocks in the top row with simulations in the bottom row. Once more, we have a reductive explanation of a complex phenomenon.

---

[26] Reynolds (1987).

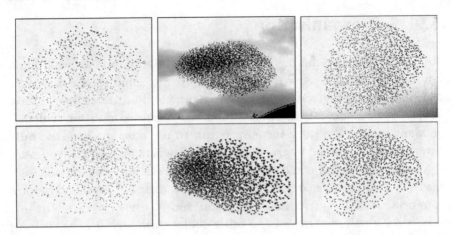

Fig. 15.2 Simulating starlings[27]

There is much interesting research on the other two examples I gave, sunflowers and snowflakes. Empirical work showing that, although they share common patterns, every sunflower and every snowflake is distinct. Theoretical work building ever more refined models. If you want to know more, see the Further Reading.

15.10   Ordinary Objects

I addressed the question of the reality of ordinary objects in Chap. 8. The most that can be said about them is that they are.

> …aspects of the world with sufficient cohesion at our scale that a group of cognitive systems with practically motivated interest in tracking them would sort them into types for book-keeping purposes.[28]

The world we live in, the world of ordinary objects, is a world of approximations. Every configuration of atoms is unique. Given our scale and cognitive faculties, we have no choice but to treat distinct objects as identical. As I argued in Sect. 8.6, this is the nature of conceptual thought. Two different balls smashing a window may look the same to us. But if you look closer, you will see that the details of the process, picosecond by picosecond, depend on the atomic make up of both ball and glass.

[27] Source: Hildenbrandt et al. (2010), reproduced by permission of Oxford University Press.
[28] Ladyman and Ross (2009), p. 5.

There's more. The fact that the laws of nature are so simple is astonishing. Everything could be dependent on everything else. Instead, something in the deep structure of the world permits ordinary objects to exist. Batterman describes what the world would be like if this wasn't the case:

> The behaviors of systems in this world would depend upon details at all spatial and temporal scales. Whether or not my coffee maker would still be a coffee maker tomorrow morning would depend sensitively upon the detailed quantum state of all the atoms in Mount Rushmore. In fact, would it even make sense to talk about systems and their behaviors? It seems there would be no real distinction between systems or behaviors at distinct scales. The very concept of behavior-at-a-scale would make no sense whatsoever. Would it even be possible to identify systems as being the same at different times and in different locations?[29]

Note this has nothing to do with a particular scale. You could replace "coffee maker" with "a water molecule" or "galaxy". In addition to the questions in the passage, we could also ask whether such a world could support complex structures such as those capable of reflecting on it. And that shows that this question should be characterised along with fine tuning arguments about why the world is the way it is. Which is the moment we step away from physics into metaphysics. Perhaps part of the sense of wonder implicit in the concept of emergence is due to this feature of the world.

15.11 Game of Life

I discussed the Game of Life in some detail in Sect. 7.6, arguing that the patterns are just useful shorthands and a reductive explanation is always available. Our brains have awe-inspiring simulation capacities. Imagine how you might run through in your mind's eye a presentation or an upcoming holiday. There, you've just simulated a simulation. However for good evolutionary reasons, we haven't got the ability to run in our head the sort of algorithm that the Game of Life is based on.

You may be lucky enough to be able to glance at a snooker table with a cue ball rolling towards an arrangement of balls, blink, and have some unconscious process in your brain give you a guess at what the final position of the balls will be. If we had a different cognitive makeup, we might be able to do the same with the Game of Life. We would then be able to look at a pattern

---

[29] Batterman (2021), p. 47.

and let some unconscious process work through the simulation then deliver a guess of the pattern at some point in the future. The former problem is no more difficult than the later, arguably it is simpler since it involves a digital rather than an analogue system. It's just that predicting trajectories was useful in our cognitive evolution, whereas predicting multiagent interactions was not.

15.12   Evolution

What to make of Stuart Kauffman's assertion of "a world beyond physics"? For me, it is a canonical example of the difference between explanation and predictability I discussed in Sect. 14.5. Perhaps one day we will understand every aspect of how living cells work, perhaps we will be able to build artificial cells based on a range of different biochemistries with a range of interesting functions. We may be able to simulate processes on the early Earth and watch the transition from chemistry to living cells happen. None of this seems unlikely to me. However, we may still not be able to answer the question of why life is the way it is. As Kauffman so forcefully argues, this may be the result of historical contingencies that only a demon, with a perfect knowledge of the past, could identify. It would be like playing pinball for 4 billion years. The causal chain is too complex to unravel.

However various studies suggests that things might not be so bad. There is evidence that evolution may be more predictable than we might think. Recent work shows that microevolution, the dynamics of a species over several generations, can predict evolution at much longer timescales.[30] And the form of current life might tell us about processes very early in the evolutionary process. A 2024 paper gives the tantalising suggestion that current biochemistry may give clues to the path from inorganic chemistry to the core of metabolism.[31]

An increasing body of evidence shows that, given conditions on the early earth, biochemistry is not an arbitrary choice but is encoded in chemistry.[32] Given a set of building blocks, and let's follow Kauffman and talk about amino acids, the evolutionary process does not conduct a random search to find the global optimum. Instead, it performs a different sort of search, guided by the problems it is trying to solve. A search which at best finds local

---

[30] Holstad et al. (2024).

[31] Goldford et al. (2024).

[32] Keller et al. (2014) discusses glycolysis and Keller et al. (2017) the citric acid cycle.

optima and which may be quite tightly constrained by environmental conditions. This means that given conditions on the early earth we would expect similar choices of proteins.

But despite all this, I suspect that Kauffman may still be right. Perhaps one day our simulations of the early earth will show only one path to living cells, and these cells will be similar to those we are made of. In other words, we may fully understand how life could have originated. But given the paucity of information about conditions on the early earth, we may never know if these simulations capture what actually happened.

15.13 Living Cells

I described chemotaxis, the ability of bacteria to sense chemical gradients and change their behaviour to swim up or down them depending on whether they represent food or a toxin. This requires two properties, top-down control and goal-directedness, which are seen by some to be characteristic of emergence.[33]

No. Instead, concepts like top-down control and goal-directedness are the result of applying the intentional stance. The behaviour of the bacterium is the result of a complex chemical mechanism. And it is a mechanism that is increasingly well understood.

Here's an explanation, adapted from a 2012 paper.[34] First look at Fig. 15.3. Starting at the top, the food molecules are shown as red diamonds. They cross the cell membrane via active pumps (the white circles). A receptor complex, composed of enzymes shown as the green torpedo like molecules, projects from the cell interior into the membrane. At the bottom of the picture, also embedded in the membrane, is the motor driving a flagellum.

The mechanism is as follows:

1. By default, the motors that drive the flagella turn counterclockwise, meaning the bacterium swims in a straight line
2. If molecules of food are present in the environment, they are transported into the cell and bind to receptors.
3. This activates the enzymatic properties of the protein, which activates a signalling chemical (shown in yellow)
4. This diffuses through the cell
5. When it reaches the flagella motors, it binds to them which causes them to change direction and the bacterium tumbles

---

[33] Winning and Bechtel (2019).
[34] Sourjik and Wingreen (2012).

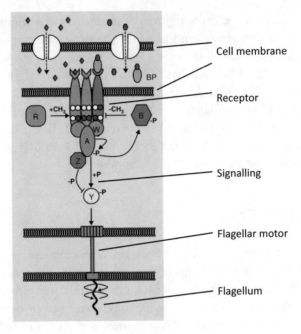

Fig. 15.3 Swim and tumble: the mechanism[35]

6. In the process, the signalling chemical is deactivated
7. When the concentration of the signalling chemical falls below a certain level, the motors revert to a clockwise direction and swimming resumes.

Part of the receptor network is a feedback control system (a fancier version of the chemical system described in Sect. 9.5) which allows it to remember past values of the gradient and compare it with the current value. If you're interested in the details, I refer you to the paper. The important point is this. There's no intention, no goal-directedness and no top-down control. There is just a set of chemical reactions.

In a recent paper with Jonathan Kopel, Ellis identifies the key problem in the relation between biology and physics is:

> …how does purpose or function emerge from purposeless physics on developmental and functional timescales?[36]

---

[35] Source: Sourjik and Wingreen (2012). https://www.sciencedirect.com/science/article/abs/pii/S09550 67411001542. Reproduced with permission from Elsevier.

[36] Ellis and Kopel (2019).

Here is the answer. *E. Coli* shows purposive behaviour, swimming towards things it likes and away from things it dislikes. But so does the wheeled mechanism I described in Sect. 13.5. Calling either purposive is a result of applying the intentional stance. Saying the bacterium swims in a particular direction in order to find food makes exactly the same mistake as saying stones fall towards the earth in order to reach their natural place.

There is no purpose in our bacterium. There are just chemical reactions. And these have been selected by evolution because they enhance the bacterium's fitness.

15.14 Turning the Page

Back in Chap. 2, I asked you to turn a page and used this as an example of mental causation and the causal power of mental states. Many see this as the strongest evidence against physicalism. Recall Ellis's assertion that emergence must be real "…because of the causal power of thoughts".[37]

But what is going on is precisely the same as my description of a bacterium in the last section. The chemical system that constitutes the bacterium senses its environment (the chemical gradient); changes its internal state (the concentration of the signalling chemical) then acts to change the world (swimming or tumbling).

In a similar way, when you read the words "turn the page" you are sensing your environment. This changes your internal state. And a mental state is no different from any other physical state, a particular arrangement of quantum fields. Then you act to change the world. Both the bacterium swimming and you turning the page are examples of the interaction of various physical systems. The different levels of complexity do not prevent this analogy from being exact.

At the end of the passage I quoted in Chap. 2, Fodor asked: "What is it, then, for a physical system to have intentional states?". The answer is that intentional states are a useful shorthand to describe the interaction of complex physical systems. Dennett didn't call it the intentional stance by accident.

The awe we feel in the face of mental causation reminds me of an anecdote recounted by mathematician Ethan Akin. He thinks he has come up with an argument to demonstrate free will, But when he explains it to one of his colleagues.

---

[37] Ellis (2016), p. 424.

He was not impressed. Shrugging, he remarked: 'I can program a computer to do that.'[38]

So far, so objective. But, you might argue that nothing here explains your subjective experience of mental causation. Partly, this is a result of our using the intentional stance to explain our own actions. Why? Because we have no direct access to the inner workings of our minds. Think of the visual system. Our apparently coherent picture of the world is stitched together through a phenomenally complex process involving approximations, predetermined categories and sheer guesswork. But this process is completely opaque. However hard we try, we can never see it happening. The best we can do is construct examples which trick the system and expose aspects of its function. As for the visual system, so for the rest of what the brain does. We need the intentional stance to explain to ourselves the results of a complex system the workings of which are hidden.

Then there is the fact that you have subjective experience at all. I will return to this in the next chapter.

15.15 Churchill's Nose

David Deutsch uses this thought experiment to show that, however much physics we know, high level concepts are essential to explain even low level facts, such as the position of that nasal atom. Leadership, war and tradition are concepts of commonsense psychology, or perhaps commonsense history. They are no more likely to be an accurate description of the world than are the concepts of commonsense physics or statistics. Deutsch's example is the intentional stance writ large. Yes, for us, it is unavoidable. But this is a function of our cognitive constitution, not the nature of the universe.

Imagine we had at our disposal a powerful AI, which knew lots about physics and lots about human history, psychology etc. Are we sure it would come up with the same explanation? For obvious reasons, we tend to see human intentions as the driving force behind the world. But Marxist historians describe humans as pawns of objective social processes. Perhaps from the alien perspective of an AI intentions and indeed consciousness might not even feature in an explanation. What arrogance to assume that our way of explaining the world is the only way.

If you are a physicalist, you may accept that the causal chain in the case of the nose is so complex, we may never be able to follow it. But this is about

---

[38] Akin (1992).

our cognitive limitations and the difficulty of predicting complex systems. The only alternative is to accept that somewhere along the path of human evolution, something which is not physics intervenes. Then you are a dualist. There is no middle way.

15.16   A Theory of Everything

You may have spent this chapter thinking something is awry. If austere physicalism is true, everything is quantum field theory. Yet in describing these examples I've barely mentioned QFT. What is going on?

An emergentist answer is given in a paper from 1999 which Laughlin coauthored with David Pines. The paper starts by arguing, as I have, that QFT is already a Theory of Everything since it is applies in all but the most extreme conditions. But the paper goes on to say:

> … the triumph of the reductionism of the Greeks is a pyrrhic victory: We have succeeded in reducing all of ordinary physical behavior to a simple, correct Theory of Everything only to discover that it has revealed exactly nothing about many things of great importance.[39]

This contention is supported by two arguments. The first relates to the computational intractability of quantum many-body problems. The combinatorial explosion involves put systems of more than a few dozen particles forever beyond classical computers. The second is what they describe as "total irrelevance" of quantum physics for a wide range of phenomena. They call such phenomena *protectorates* defined as:

> …a stable state of matter whose generic low-energy properties are determined by a higher organizing principle and nothing else.[40]

I'm going to give four arguments to counter Laughlin and Pines's case. First, it's important to note that they give no proof of their assertion and that neither can I prove that everything can be derived from QFT. So the discussion can only be about plausibility.

Computational limitations: Pines and Laughlin's discussion does not mention quantum computers. If we can build a universal quantum computer, the Church-Turing-Deutsch principle means we can simulate every system.

---

[39] Laughlin and Pines (2000).
[40] Laughlin and Pines (2000).

Universality: protectorates are another term for universality. We've seen that for phase transitions, there is no such thing as strict universality. Transitions are continuous and in general depend on the microscopic details of the system. I speculate that the same is true for all other protectorates. We know the fractional quantum Hall effect is accurate down to one part in a billion. But if we measured temperature in human-scale objects only to that accuracy, it would also be termed a protectorate. The empirical question of the discontinuity of topological transitions remains open.

State of knowledge: we could call nucleons protectorates since their properties are fixed even though the tangle of virtual particles that comprises them is constantly changing. But QCD explains why this is so. We could likewise call boxes of gas protectorates, but statistical physics explains why their properties are broadly independent of their precise compositions. It starts sounding as if calling something a protectorate is simply a way of saying we haven't worked it out yet. An important method to gain insight into the mechanism of such phenomena is by using simple models. In the same way as the Ising model clarifies phase transitions, minimal models, such as the one of the fractional quantum Hall effect I mentioned in Sect. 15.7, allow us to look inside protectorates and see how they are constituted.

Continuity: where do the "higher organising principles" come from? As a thought experiment, let's imagine building atom by atom a material in which the fractional quantum Hall effect is observed. Start with one atom; add another then another in such a way that they form part of the structure of the material. Clearly the behaviour of these few atoms is dependent on quantum physics. What happens as we keep adding in atoms? Remember how the Ising model shows how a phase transitions gets sharper and sharper as the number of spins increases? I would speculate that the same is true for all such phenomena: the quantum Hall effect plateaux gradually sharpen as more and more atoms become entangled in the wave function of the system as a whole. If this isn't the case, then there must be some N such that for N atoms things proceed according to QFT and for N + 1 "higher organizing principles" take over. Of course, there's no logical reason this couldn't be the case but it would be unlike anything in physics.

You might have noticed that the first and last of my arguments are linked. The process I asked you to imagine is exactly the sort of controlled experiment a quantum simulation would permit. A whole flurry of recent papers use quantum simulations to investigate many-body problems that would be intractable to a classical computer. One models a quantum 2D Ising model with 300 sites[41] allowing the details of the coupling between sites to be precisely tuned. Another studies he effects of long-range entanglement in a chain of 51 sites.[42] Such techniques can in principle be applied to the full range of phenomena which motivate Laughlin and Pines's argument and may resolve the question of the applicability of the Theory of Everything.

15.17   The Many-Body Challenge

Do you still think non-reductive physicalism is possible? If so, the thought experiment of the previous section can be turned into a way of demonstrating that the arguments I've made to the contrary are wrong.

1. Specify your system
2. Write down a many-body description of your system making whatever approximations you need.
3. Explain why your "emergent" property will not decrease gradually as you remove particles one by one.

Of course it would be best to do this formally, as in the model of Debye screening discussed in Sect. 9.3. But an explanation in words would be a good start.

When does your "emergent" property disappear[43]? Is it sudden? If so, what changes between N + 1 particles and N? If it's gradual, why can't I measure it down at the level of atoms? Remember, what matters is that something new emerges which cannot be explained in terms of the interactions of its components. It seems to be that if such an explanation exists, it must necessarily involve new physics and hence be strong emergence. Prove me wrong.

15.18   Further Reading

The Further Reading for Chap. 2 contained suggestions for each of the examples.

An accessible discussion of quantum chemistry can be found at Arturo Robertazzi's blog, https://www.arturorobertazzi.it/ and a discussion of the

---

[41] Guo et al. (2024).

[42] Joshi et al. (2023).

[43] This will of course be above some minimum. It makes no sense to talk of a 3-body gravitational interaction with less than three bodies!

relation between quantum computing and quantum chemistry in Lanyon et al. (2010). For a standard textbook derivation of the fluid dynamics of convection, see Tritton (2007), Chap. 14. Falkenburg (2007) is a fascinating discussion of the place of particles in physics. For a crystal clear explanation of the empirical work and theoretical models behind the quantum Hall effects, see Stormer (1999).

The bible of snowflake science is Libbrecht (2022) and an interesting theoretical model is in Demange et al. (2017). As for sunflowers, empirical work is in Swinton et al. (2016) and a model in Mirabet et al. (2012).

More suggestions for reading can be found at www.TheMaterialWorld.net.

References

Akin E (1992) The spiteful computer: A determinism paradox. The Mathematical Intelligencer 14:45–47. https://doi.org/10.1007/BF03025213

Batterman RW (2021) A middle way: a non-fundamental approach to many-body physics. Oxford University Press, New York, NY

Bishop RC (2019) The physics of emergence. Morgan & Claypool Publishers, San Rafael, CA

Bishop RC, Silberstein M, Pexton M (2022) Emergence in context: a treatise in twenty-first century natural philosophy. Oxford University Press, Oxford

Bobylev AV (2018) Boltzmann equation and hydrodynamics beyond Navier–Stokes. Phil Trans R Soc A 376:20170227. https://doi.org/10.1098/rsta.2017.0227

Burgess CP (2004) Quantum Gravity in Everyday Life: General Relativity as an Effective Field Theory. Living Rev Relativ 7:5. https://doi.org/10.12942/lrr-2004-5

Chen G (2021) Non-Fourier phonon heat conduction at the microscale and nanoscale. Nat Rev Phys 3:555–569. https://doi.org/10.1038/s42254-021-00334-1

Chen X, Gu Z-C, Wen X-G (2010) Local unitary transformation, long-range quantum entanglement, wave function renormalization, and topological order. Phys Rev B 82:155138. https://doi.org/10.1103/PhysRevB.82.155138

Demange G, Zapolsky H, Patte R, Brunel M (2017) A phase field model for snow crystal growth in three dimensions. npj Comput Mater 3:15. https://doi.org/10.1038/s41524-017-0015-1

Dürr S, Fodor Z, Frison J, et al (2008) Ab Initio Determination of Light Hadron Masses. Science 322:1224–1227.https://doi.org/10.1126/science.1163233

Einstein A, Podolsky B, Rosen N (1935) Can Quantum-Mechanical Description of Physical Reality Be Considered Complete? Phys Rev 47:777–780.https://doi.org/10.1103/PhysRev.47.777

Ellis G (2016) How can physics underlie the mind? Springer, Berlin

Ellis G (2020) Emergence in Solid State Physics and Biology. Found Phys 50:1098–1139.https://doi.org/10.1007/s10701-020-00367-z

Ellis G, Kopel J (2019) The Dynamical Emergence of Biology From Physics: Branching Causation via Biomolecules. Frontiers in Physiology 9. https://doi.org/10.3389/fphys.2018.01966

Falkenburg B (2007) Particle metaphysics: a critical account of subatomic reality. Springer, Berlin ; New York

Fisher ME (2016) Condensed Matter Physics: Does Quantum Mechanics Matter. In: Excursions in the Land of Statistical Physics. World Scientific

Goldford JE, Smith HB, Longo LM, et al (2024) Primitive purine biosynthesis connects ancient geochemistry to modern metabolism. Nat Ecol Evol 8:999–1009. https://doi.org/10.1038/s41559-024-02361-4

Guo S-A, Wu Y-K, Ye J, et al (2024) A site-resolved two-dimensional quantum simulator with hundreds of trapped ions. Nature. https://doi.org/10.1038/s41586-024-07459-0

Heitler W, London F (1927) Wechselwirkung neutraler Atome und homoeopolare Bindung nach der Quantenmechanik. Z Physik 44:455–472. https://doi.org/10.1007/BF01397394

Hildenbrandt H, Carere C, Hemelrijk CK (2010) Self-organized aerial displays of thousands of starlings: a model. Behavioral Ecology 21:1349–1359.https://doi.org/10.1093/beheco/arq149

Holstad A, Voje KL, Opedal ØH, et al (2024) Evolvability predicts macroevolution under fluctuating selection. Science 384:688–693.https://doi.org/10.1126/science.adi8722

Joshi MK, Kokail C, van Bijnen R, et al (2023) Exploring large-scale entanglement in quantum simulation. Nature 624:539–544.https://doi.org/10.1038/s41586-023-06768-0

Kadanoff LP (2009) More is the Same; Phase Transitions and Mean Field Theories. J Stat Phys 137:777–797. https://doi.org/10.1007/s10955-009-9814-1

Keller MA, Kampjut D, Harrison SA, Ralser M (2017) Sulfate radicals enable a non-enzymatic Krebs cycle precursor. Nat Ecol Evol 1:0083. https://doi.org/10.1038/s41559-017-0083

Keller MA, Turchyn AV, Ralser M (2014) Non-enzymatic glycolysis and pentose phosphate pathway-like reactions in a plausible A rchean ocean. Molecular Systems Biology 10:725.https://doi.org/10.1002/msb.20145228

Kleftogiannis I, Amanatidis I (2021) Fractional-quantum-Hall-effect (FQHE) in 1D Hubbard models. Eur Phys J B 94:41. https://doi.org/10.1140/epjb/s10051-021-00050-w

Ladyman J, Ross D (2009) Every thing must go: metaphysics naturalized. Oxford University Press, Oxford

Lanyon BP, Whitfield JD, Gillett GG, et al (2010) Towards quantum chemistry on a quantum computer. Nature Chem 2:106–111.https://doi.org/10.1038/nchem.483

Laughlin RB, Pines D (2000) The Theory of Everything. Proc Natl Acad Sci USA 97:28–31. https://doi.org/10.1073/pnas.97.1.28

Libbrecht K (2022) Snow crystals: a case study in spontaneous structure formation. Princeton University Press, Princeton, NJ

Mirabet V, Besnard F, Vernoux T, Boudaoud A (2012) Noise and Robustness in Phyllotaxis. PLoS Comput Biol 8:e1002389. https://doi.org/10.1371/journal.pcbi.1002389

Ng RC, El Sachat A, Cespedes F, et al (2022) Excitation and detection of acoustic phonons in nanoscale systems. Nanoscale 14:13428–13451. https://doi.org/10.1039/D2NR04100F

Prigogine I, Stengers I (1984) Order out of chaos: man's new dialogue with nature. Bantam Books, Toronto; New York, N.Y

Primas H (1998) Emergence in exact natural science. Acta Polytechnica Scandinavica Mathematics and Computing Series 91. https://philsci-archive.pitt.edu/953/

Ramakrishnan R, Dral PO, Rupp M, von Lilienfeld OA (2014) Quantum chemistry structures and properties of 134 kilo molecules. Sci Data 1:140022. https://doi.org/10.1038/sdata.2014.22

Rapaport DC (1988) Molecular-Dynamics Study of Rayleigh-Bénard Convection. Phys Rev Lett 60:2480–2483. https://doi.org/10.1103/PhysRevLett.60.2480

Reynolds CW (1987) Flocks, herds and schools: A distributed behavioral model. In: Proceedings of the 14th annual conference on Computer graphics and interactive techniques. ACM, pp 25–34

Sourjik V, Wingreen NS (2012) Responding to chemical gradients: bacterial chemotaxis. Current Opinion in Cell Biology 24:262–268. https://doi.org/10.1016/j.ceb.2011.11.008

Stormer HL (1999) Nobel Lecture: The fractional quantum Hall effect. Rev Mod Phys 71:875–889.https://doi.org/10.1103/RevModPhys.71.875

Swinton J, Ochu E, The MSI Turing's Sunflower Consortium (2016) Novel Fibonacci and non-Fibonacci structure in the sunflower: results of a citizen science experiment. R Soc open sci 3:160091. https://doi.org/10.1098/rsos.160091

Tritton DJ (2007) Physical fluid dynamics. Clarendon Press, Oxford

Watanabe T, Kaburaki H (1997) Particle simulation of three-dimensional convection patterns in a Rayleigh-Bénard system. Phys Rev E 56:1218–1221. https://doi.org/10.1103/PhysRevE.56.1218

Winning J, Bechtel W (2019) Being emergence vs. pattern emergence. In: Gibb SC (ed) The Routledge handbook of emergence. Routledge, New York

16

Epilogue

Summary Free will and consciousness have been mostly absent from the preceding chapters. Some see them as central. Indeed, David Chalmers argues that consciousness is the only example of strong emergence. In this chapter, I will use some thought experiments to clarify the issues. My aim is to show that there is no reason they cannot be incorporated in a physicalist account.

You are not authoritative about what is happening to you, only about what seems to be happening to you[1]

Free will and consciousness are often seen as emergent phenomena par excellence. But so far I've mostly avoided discussing them. This is largely because they are among the few phenomena for which we are entirely reliant on first-person, subjective data. This puts them outside the scope of scientific enquiry. If the only refuge left for emergence is your impression of what goes on in your head, I am happy to rest my case.

However, I believe a physicalist account of free will and consciousness is possible. Producing one is a substantial project. For the moment, I would like to give you a few pointers to suggest that, while we are a long way from understanding these phenomena, they are nothing for a physicalist to fear.

---

[1] Dennett (1991), p. 96.

© The Author(s), under exclusive license to Springer Nature
Switzerland AG 2025
L. Graham, *Physics Fixes All the Facts*, The Frontiers Collection,
https://doi.org/10.1007/978-3-031-69288-8_16

16.1 Free Will

We choose, act and in so doing change the world. Yet so does a bacterium when it stops tumbling and starts swimming. And so does a robot lawnmower when it turns to the left rather than the right. The bacterium follows an "If... then" rule encoded in a network of chemical reactions. The lawnmower follows an "if... then" rule encoded in a network of transistors. There is some randomness in bacterial behaviour due to the molecular storm. A robot may include a pseudo-random number generator. This would make their behaviour more difficult to understand but has nothing to do with freedom. We are free in precisely the same sense as a bacterium or a lawnmower.

The concept of free well is such an incoherent mess that it makes emergence look well defined. All we know is that it's obvious to everyone that we have it. In the same way it's obvious that the earth is flat and doesn't move.

Our impression of free will is another instance of the intentional stance (Sect. 13.5). A key part of our cognitive evolution is driven by the need to navigate the complex environment of small social groups. To do this, we need to evolve simple ways of interpreting the complex systems, our fellow humans, which we interact with. A fast and frugal heuristic to do this is to treat them as agents freely pursuing their desires and beliefs. Using it means we don't need to worry about their hidden internal structure. The same model can be applied to ants, cats or machines. This faculty is often called mindreading. It involves a projection of our vastly simplified models onto the world.

In attempting to understand our own behaviour, we face exactly the same problem. Almost all of what we do is carried out by unconscious mechanisms which are completely untransparent. Try to see into how you move your arm, or how your perceptive system constructs the image you have in front of you. These systems are encapsulated and inaccessible to us in exactly the same way that the inner workings of other creatures are inaccessible.

To make sense of all this, we apply the intentional stance to ourselves. Not only do we treat other systems as agents, we treat ourselves as agents. What's more, the interpretative process itself is also untransparent. We experience ourselves or others as having emotional states, being happy, impatient or angry, without realising that these are interpretations. They are psychological coarse graining, simplified representations of a hugely complex system most of which is inaccessible to us.

We have a narrative mode of consciousness which gives regular reports on our internal state. Part of its content are outputs of the mindreading module. Some thinkers see this facility as one of the things which differentiates humans from primates and a key step in the development of morality.[2] However it is not a reliable faculty. Here is Daniel Dennett's description of its function:

> ...we are all virtuoso novelists, who find ourselves engaged in all sorts of behavior, more or less unified, but sometimes disunified, and we always put the best 'faces' on it we can. We try to make all of our material cohere into a single good story.[3]

The attempt to invent a story good enough to explain our actions to ourselves and to others is also known as *confabulation*. In combination with the intentional stance, it explains the origin of our subjective impression of free will. As an illustration, here's a light-hearted excursion into science fiction.

Disclaimer: Any resemblance between this story and actual brain function is pure coincidence.

Churchland Systems of San Diego have long been at the forefront of developments in full-brain simulation technology. Their first consumer product is Elmat™ which makes high level brain processes accessible in ordinary language and in close to real-time. I was invited to join the beta test program.

Preparation included a brain scan at sub-neuronal resolution, a longer version of those everybody has as part of their yearly check-up of cognitive function. Then came the injection of a suite of cerebral nanosensors. Again, these are an adaptation of technology that is regularly used to repair brain damage. Finally, a long and tedious process of calibration as their AI took me through a whole range of physical and cognitive exercises. Once this was done, I was issued with the brain monitor which uploads signals from the nanosensors into the quantum cloud.

Here's what happened the first time I used it. I had been working since lunchtime and was feeling peckish. I stretched, figured 3 pm was a good time for a snack so stood up and wandered downstairs in search of an apple. In the kitchen, I thought "What the heck, I deserve a treat" and gobbled a whole tub of ice cream.

A few minutes later, disappointed by my lack of willpower, I launched the Elmat™ app to see what had been going on in my brain. For the modular

---

[2] Gibbard (1992).
[3] Dennett (1992).

event report, I unchecked the boxes relating to the dull stuff (perception, homeostasis, detailed motor plans, etc.) to focus on what lay behind my decision. I also requested the confabulation report which gives an idea of the relation between my explanation and what actually took place in my brain. Here are the reports.

Modular Events and Relevant Conscious Activity (RCA)

Time		RCA (%)
14:30	Hunger signals sent by digestive system/blood sugar monitor	0
14:52:10	Signal intensity crosses a threshold. The body monitoring module creates a desire for food which is then passed to the desire selection module	0
	In the desire selection module, it enters a competition with other current desires: carry on working, go for a walk, check the news	0
	Other modules allocate weights to these desires; getting food acquires the highest weight so wins the competition	0
14:52:11	This is broadcast to all other modules	0
	A memory search is carried out to see what food is available and where, resulting in a list of possible actions	0
	These actions are passed to the action selection module and another competition starts, with the various options attracting weights from other modules. At the end, two possibilities have roughly equal weights: ice cream or apple	0
	Each is passed to the mental rehearsal module which simulates the process of acquiring the food and the results of eating it. During the simulation other brain modules behave as they would if the action were actually carried out	0
	The ice cream option picks up weight from the fuelling module due to its fat and calorific content. The self-control module offsets this with a negative weighting	0
	The prospection facility is consulted and suggests that proximity to the food will increase the weight from the fuelling module	0
14:52:12	The action that is chosen is 'go to the kitchen and replan in front of the fridge'. An appropriate motor action plan is retrieved from memory	0
	The motor plan is started. Its first step is to prepare for movement by stretching	0
14:52:15	The access consciousness process registers the physical movement	5

(continued)

(continued)

Time		RCA (%)
	The rest of the motor plan is put into action	0
15:02	In front of the fridge, the two options are again passed to the action selection module. The proximity of the food increases the weight given to the fuelling module. The self-control module is fatigued after its continuous use during the preceding hours of work so attracts less weight. This means ice cream wins the competition	0
	Memory is consulted for a motor plan to find and eat the ice cream	0
15:03	The plan is executed and the ice cream is eaten. With the plan running, neither the self-control nor the fuelling module play any further role in the process	0
15:10:28	A request for an explanation is passed from the consciousness process to the mindreading module	10
	The mindreading module searches memory for relevant information. The only events it finds are: • Stretching at 15:00 • Eating the ice cream at 15:03	0
	The mindreading module evolved to explain the behaviour of other people and was then adapted to provide explanations of self. Given the two events in memory it: • Fills in the gaps using memories from similar past situations • Uses this information to estimate the time at which the action started • Interprets the memories using the commonsense psychological concepts of desire, choice and justification	0
15:11:29	The result is passed to the language module. A motor plan is selected to type it out	10

Confabulation report

Between the start of the report at 14:30 and the end of the action at 15:03 the narrative consciousness process was inactive. The access consciousness process was involved in monitoring various concurrent background processes, among them an attempt to solve a mathematical problem, aesthetic appreciation of the blue sky and planning the evening's meal.

Confabulation index: 93%.

(a value about 80% implies the mindreading module's output is ex-post justification and bears minimal relation to actual brain events).

User Notes

Please remember Elmat is currently in beta-test. There are a number of issues and restrictions,

- Reliability is high but not 100%. We believe this is mostly due to measurement limitations. Computing is easy. Measuring the physical world is more difficult. However, there is a residual effect which we only observe in humans and not animal subjects. We call this the mischievous module. Our preliminary research suggests a poorly understood process gives extra weight to actions which confound our system. The few remaining dualists describe this with the anachronism free will.
- Future versions will add the possibility to drill down from the high level reports to a range of further levels, presented with cutting-edge 3D visualisation techniques

  - Modular activity: the behaviour of different modular components of the brain
  - Neuronal activity: the behaviour and interaction between neurons in terms of the theory of neurotronics.

- Research versions of our system can go further. However the resulting data is so rich as to be incomprehensible to even the most advanced AIs.

  - The high level concepts of neurotronics are described in terms of the behaviour of individual neurons
  - The behaviour of neurons can be described terms of the interaction of their constituent molecules
  - From there, it's physics all the way down

16.2   The "Hard" Problem of Consciousness

David Chalmers claims that phenomenal consciousness is a problem different from all others. There are many aspects to consciousness, for example the narrative mode I mentioned in the previous section. All these are potentially accessible to neuroscience. But an explanation of why we have subjective experience at all, why it is "like something" to be us, will remain forever beyond science.

Chalmers sees phenomenal consciousness as the only example of strong emergence:

… given a complete catalogue of physical facts about the world, supplemented by a complete catalogue of facts about consciousness, a Laplacean super-being could, in principle, deduce all the high level facts about the world, including the high level facts about chemistry, biology, economics, and so on.[4]

For Chalmers, physics does not fix all the facts, only those not to do with consciousness. Since his original 1995 paper,[5] the idea has been hugely influential. What bothers me most about it is the whiff of anthropocentrism. We may no longer be at the centre of things, but we are still the hardest problem in the universe. How cool is that?

Chalmers is correct that it is a problem like no other. But this is because, so far at least, we have only first-person data. I have no way of knowing whether your subjective experience is the same as mine. I have no way of knowing whether you have any subjective experience at all. Without objective evidence, the question is outside the scope of science.

Until we have such evidence, my take is that everyone needs to calm down. While Chalmers's dualism is logically coherent (and when we have evidence it may prove him correct), it seems odd to base such huge metaphysical consequences on such flimsy ground. It makes me think of Penrose's assertion that brains are hypercomputers (Box 5.4). What is it about consciousness that leads to such dramatic assertions? Is it that phenomenal consciousness is as resistant to clear definition as free will? And like free will, we are all utterly certain that we have it?

Consciousness plays a much smaller role in our lives than we might think. Almost everything we do happens at a level completely inaccessible to us. Our knowledge of these processes comes either from observing our actions or from the confabulated accounts of narrative consciousness. Here's Daniel Dennett writing about Fodor:

> [he] once made the point with the aid of an amusing confession: he acknowledged that when he was thinking his hardest, the only sort of linguistic items he was conscious of were snatches along the lines of 'C'mon, Jerry, you can do it!'[6]

Chalmers defends his argument with a number of thought experiments. I'm going to discuss two of them. The first is known as the zombie argument[7] A zombie is an entity who is exactly the same as a human in all

---

[4] Chalmers (2006).
[5] Chalmers (1995).
[6] Dennett (1991), p. 303.
[7] Kirk (2023).

physical respects. But they do not have subjective experience, they are dark inside. If zombies are possible, subjective experience is something in addition to physical structure. Therefore physicalism is false.

The second is known as the knowledge argument.[8] Imagine a scientist, call them M, who has no colour vision. They study the colour red from all possible perspectives, physical, optical, neuroscientific etc. until they know everything that is to be known about red. Despite knowing all the physical facts, they will know nothing about the subjective experience of red. Therefore subjective experience is not a physical fact and physicalism is false.

Let me address these with my own thought experiment. After decades of work, we finally understand consciousness. To demonstrate our understanding, we regularly build artificial consciousnesses (ACs) which we can manipulate in anyway we want.[9]

What would this imply for the two arguments? Since consciousness would be understood as a physical phenomenon, the problem of zombies immediately evaporates. Zombies cannot exist and our ACs will show us precisely why. If we think zombies are conceivable, this is no more than yet another failure of our intuition to tell us something meaningful about the world, another example of the mind projection fallacy.

As for M, let's extend their knowledge to include everything that we would have learnt about consciousness from the ACs. They now deeply understand the nature of subjective experience. They understand precisely the neural mechanisms which give rise to it. They can observe step by step the subjective experiences of various ACs when presented with the colour red. In the language of Sect. 4.4, they have full representational understanding.

The knowledge argument then says no more than that M lacks imaginative understanding. In exactly the same way that we lack imaginative understanding of quantum physics. It would be no big deal. A more capable entity would be able to take all the information, use it to run an internal simulation and experience red without ever seeing it.

There is something self-referential about the "hard" problem. If we assume it exists, then these and other thought experiments confirm it. However, if we assume that a physical explanation of consciousness is possible, the difficulties evaporate. Of course, without a physicalist account of consciousness, these arguments are no more than words.

---

[8] Nida-Rümelin and O Conaill (2024).

[9] Such a possibility has fearsome ethical implications. See Metzinger (2009), Chap. 9 and Chalmers (2022), Chap. 18.

16.3 The Meaning of Life

The intentional stance leads us to see a world full of purposes. From the narrow point of view that it provides, the purpose of some creature appears to be to find food so it can reproduce and perpetuate its species. All a demon would see is a quantum field evolving towards maximum entropy subject to complex constraints in some high-dimensional space.

We can say a bit more. The rate at which a system increases the entropy of its surroundings is effectively the rate at which it dissipates energy. To get a handle on this, let's do some back of the envelope calculations. The production of entropy is, essentially, the same as the dissipation of heat. To calculate the rate of heat dissipation per unit mass of a system, all we need to know is its power output and mass.

The sun: it generates around 4×10^{26} W and its mass is 2×10^{30} kg which gives a power per unit mass of 2×10^{-4} W/kg.

A human: let's take a person weighing 65 kg and consuming 2000 calories per day. This gives a dissipation rate of 1.5 W/kg. A kilogram of human dissipates 7000 times as much energy as a kilogram of sun.

A computer: the laptop I am using weighs around 1 kg and has a power adaptor rated at 45 W so dissipates 45 W/kg, 30 times as much as a human.

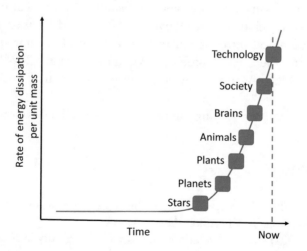

Fig. 16.1 Dissipation in the universe[10]

---

[10] Redrawn from Chaisson (2002), Fig. 28.

Figure 16.1 shows a cartoon of the result of extending this calculation to more systems, with the rate of energy dissipation on the y-axis against the age of the universe on the x. The message of the graph is striking. As the universe gets older, structures develop which dissipate energy at an ever-greater rate. There are a few hints as to the physical mechanism behind this, but as yet no general explanation.

The process of evolution itself can be seen in this light. Can you think of a better way of dissipating energy than putting loads of effort into building increasingly complex physical systems then letting them eat each other? Of course you can. Have some of those complex physical systems use their imperfect knowledge of the world to build further systems, call them technology, which dissipate energy at far higher rates.

Remember, the universe will most likely end in a big freeze when there is no further possibility for energy dissipation. Then Fig. 16.1 suggests that, as the universe ages, it stumbles across systems which bring it ever closer to its end. Humans and their products are, at least in this corner of the universe, currently the best way of doing this. If you're looking for an external meaning of life, this is it.

16.4   Further Reading

For an excellent discussion of free will, see Sapolsky (2023). For two different approaches to consciousness, see Chalmers (1997) and Dennett (1991). For a more philosophical introduction to free will, see O'Connor and Franklin (2022) and to the "hard" problem Wu and Morales (2024), Sec. 1.5. Carruthers (2006) is a superb introduction to the modular workings of the brain.

More suggestions for reading can be found at www.TheMaterialWorld.net.

References

Carruthers P (2006) The architecture of the mind: massive modularity and the flexibility of thought. Clarendon Press, Oxford

Chaisson EJ (2002) Cosmic Evolution: The Rise of Complexity in Nature. Harvard University Press

Chalmers DJ (1995) Facing Up to the Problem of Consciousness. Journal of Consciousness Studies 2:200–19, https://doi.org/10.1093/acprof:oso/978019531 1105.003.0001

Chalmers DJ (1997) The Conscious Mind In Search of a Fundamental Theory. Oxford University Press, Oxford

Chalmers DJ (2006) Strong and Weak Emergence. In: Davies P, Clayton P (eds) The Re-Emergence of Emergence: The Emergentist Hypothesis From Science to Religion. Oxford University Press

Chalmers DJ (2022) Reality+: virtual worlds and the problems of philosophy. W. W. Norton & Company, New York (N.Y.)

Dennett DC (1991) Consciousness explained. Little, Brown, Boston

Dennett DC (1992) The Self as a Center of Narrative Gravity. In: Kessel FS, Cole PM, Johnson DL, University of Houston (eds) Self and consciousness: multiple perspectives. L. Erlbaum, Hillsdale, N.J

Gibbard A (1992) Wise choices, apt feelings: a theory of normative judgment. Harvard University Press, Cambridge, Mass

Kirk R (2023) Zombies. The Stanford Encyclopedia of Philosophy. https://plato.sta nford.edu/entries/zombies/

Metzinger T (2009) The ego tunnel: the science of the mind and the myth of the self. Basic Books, New York

Nida-Rümelin M, O Conaill D (2024) Qualia: The Knowledge Argument. The Stanford Encyclopedia of Philosophy. https://plato.stanford.edu/entries/qualia-knowledge/

O'Connor T, Franklin C (2022) Free Will. The Stanford Encyclopedia of Philosophy. https://plato.stanford.edu/entries/freewill/

Sapolsky RM (2023) Determined: a science of life without free will. Penguin Press, New York

Wu W, Morales J (2024) The Neuroscience of Consciousness. The Stanford Encyclopedia of Philosophy. https://plato.stanford.edu/entries/consciousness-neuroscience/

Appendix: Supporting Material

The first section of this Appendix lists definitions of emergence. The remainder includes more technical material. While I've kept the body of this book free of maths, I sometimes refer to results which can be best explained with a few lines of algebra.

A.1. Emergence: 75 Definitions

This section collects definitions of emergence that I came across in my reading. It is composed of citations without any commentary. Where an author gives different definitions of the same idea, I've chosen the one that seems clearest to me. Please bear in mind that extracting a short passage risks misrepresenting a writer's intention.

The list makes no attempt to be exhaustive. Indeed Ernst Mandelbaum argues that the concept of emergence goes further into the philosophical tradition than I have any desire to venture:

> One can find idealists such as Hegel, materialists such as Marx and Engels, positivists such as Comte, non-dualists such as Alexander and R. W. Sellars, and dualists such as Lovejoy and Broad, all holding doctrines of emergence which (with the exception of Hegel's) were remarkably similar.[1]

---

[1] Mandelbaum (1974), p. 380.

© The Editor(s) (if applicable) and The Author(s), under exclusive
license to Springer Nature Switzerland AG 2025
L. Graham, *Physics Fixes All the Facts*, The Frontiers Collection,
https://doi.org/10.1007/978-3-031-69288-8

1. Being Emergence (Winning and Bechtel)

"The being of X: If the ontological category of X is C, then the being of X is whatever it is about X that allows it to count as an instance of C.

Y is being-dependent on X = def: Y's counting as an instance of ontological category C1 is dependent on X's counting as an instance of ontological category C2 (for some C1 and C2).

If X is being-dependent on Y but X does not reduce to Y, then X emerges from Y."[2]

2. Causal Emergence (Sartenaer)

"A property E causally emerges from an underlying physical basis $\{B_i\}$ [if and only if] (1) E supervenes on—but is not realized in—$\{B_i\}$ [substantial continuity] and (2) E downwardly acts—in Sperry's sense—on $\{B_i\}$ [causal discontinuity]."

where.

"Sperry-style downward causation—a downwardly causal relation that is efficient, reflexive and diachronic"[3]

3. Causal Emergence (Searle)

"… system features cannot be figured out just from the composition of the elements and environmental relations; they have to be explained in terms of the causal interactions among the elements. Let's call these 'causally emergent system features'. Solidity, liquidity, and transparency are examples of causally emergent system features."[4]

4. Coarse-Grained Emergence (Palacios)

"A coarse-grained description of a system emerges synchronically upon a fine-grained description, [if and only if] the former has terms denoting properties or behavior that are novel and autonomous with respect to the latter, and these properties or behavior supervene upon the behavior of the components of the fine-grained description."[5]

---

[2] Winning and Bechtel (2019).
[3] Sartenaer (2016).
[4] Searle (2008), p. 69.
[5] Palacios (2022), p. 39.

5. Computational Emergence (Cariani)

"...complex global forms can arise from local computational interactions... Because its ontology only admits of micro-deterministic computational interactions... there is a strong platonic component to the world-view: the ideal forms of computational behaviours can be abstracted completely from their material substrates, and the material world can be left completely for a virtual one."[6]

6. Conceptual Emergence (Humphreys)

"...an entity, such as a state or a property, is conceptually emergent with respect to theoretical framework F if and only if a conceptual or descriptive apparatus that is not in F must be developed in order to effectively represent that entity."[7]

7. Contextual Emergence (Bishop, Silberstein and Pexton)

"Properties and behaviors in a particular domain (including its laws) at best offer some necessary but no sufficient conditions to determine properties and behaviors in another domain."[8]

8. Deducible or Computational Emergence (Baas and Emmeche)

"There exists a deduction or computational process or theory D such that $P \in Obs^2(S^2)$ can be determined by D from $(S^{1:i}, Obs^1, Int^1)$".[9] For a definition of the terms see Emergence (Baas and Emmeche).

9. Degrees of Freedom (DOF) Emergence (Wilson)

"An entity E is weakly emergent from some entities e_i if
  1. E is composed of the e_i, as a result of imposing some constraint(s) on the e_i.
  2. For some characteristic state S of E: at least one of the DOF required to characterize a realizing system of E (consisting of the e_i standing in the

---

[6] Cariani (1991).
[7] Humphreys (2008).
[8] Bishop et al. (2022), p. 27.
[9] Baas and Emmeche (1997).

e_i-level relations relevant to composing E) as being in S is eliminated from the DOF required to characterize E as being in S.

3. For every characteristic state S of E: Every reduction, restriction, or elimination in the DOF needed to characterize E as being in S is associated with e_i-level constraints.

4. The law-governed properties and behavior of E are completely determined by the law-governed properties and behavior of the e_i, when the e_i stand in the e_i-level relations relevant to their composing E."[10]

10. Diachronic and Synchronic emergence (Rueger)

"...the strategy of characterizing the relation of higher level to lower level properties as diachronic or synchronic emergence is the same. In the diachronic case we simply compare the behavioural properties of the system at a time (lower level) with those at a later time (higher level). In the synchronic case we decompose the system (or, rather, its behaviour: the higher level) into a combination of lower level sub-systems (or, rather, the behaviour generated by them) which are identified through a perturbation analysis of the full system. In each case, the higher level properties are said to be emergent if they are 'novel' or 'irreducible' with respect to the lower level properties."[11]

11. Diachronic Emergence (Sartenaer)

"Diachronic emergence is an empirical relation between an emergent and its emergence basis such that (a) the emergent is causally determined by its emergence basis, and (b) it is not possible to trace the causal chain that goes from the emergence basis to the emergent."[12]

12. Diachronic Structure Emergentism (Stephan)

"...[is] characterized by the thesis of structure unpredictability. This version of unpredictability had no role to play in the classical literature on emergence, but it gains increasing importance to phenomena studied in robotics and A-Life research:

---

[10] Wilson (2021).

[11] Rueger (2000).

[12] Sartenaer (2015).

Structure unpredictability. The rise of a novel structure is unpredictable, in principle, if its formation is governed by laws of deterministic chaos. Likewise, any property that is instantiated by the novel structure is unpredictable, in principle."[13]

13. Diachronic Transformational Emergence (Guay and Sartenaer)

"… let us consider a natural system S at two successive times t_1 and t_2 of its evolution. One will say—and in this lies the general, metaphysical account of [TE]—that the given system at t_2 (S_2) transformationally emerges from the same system at t_1 (S_1) if and only if there exists a transformation [Tr] such that:

- S_2 is the product of a spatiotemporally continuous process going from S_1 (for example causal, and possibly fully deterministic). In particular, the "realm" R to which S_1 and S_2 commonly belong (e.g. the physical realm) is closed, to the effect that nothing outside of R participates in S_1 bringing about S_2. And yet:
- S_2 exhibits new entities, properties or powers that do not exist in S_1, and that are furthermore forbidden to exist in S_1 according to the laws $\{L_1^i\}_{i=1}^n$ governing S_1. Accordingly, different laws $\{L_2^i\}_{i=1}^m$ govern S_2."[14]

14. Emergence (Alexander)

"The emergence of a new quality from any level of existence means that at that level there comes into being a certain constellation or collocation of the motions belonging to that level, and possessing the quality appropriate to it, and this collocation possesses a new quality distinctive of the higher complex. The quality and the constellation to which it belongs are at once new and expressible without residue in terms of the processes proper to the level from which they emerge… The higher quality emerges from the lower level of existence and has its roots therein, but it emerges therefrom, and it does not belong to that lower level, but constitutes its possessor a new order of existent with its special laws of behaviour. The existence of emergent qualities thus

---

[13] Stephan (2006).
[14] Guay and Sartenaer (2016), p. 303.

described is something to be noted, as some would say, under the compulsion of brute empirical fact... It admits no explanation."[15]

15. Emergence (Aristotle)

"For of all things that have several parts and where the totality of them is not like a heap, but the whole is something beyond the parts, there is some cause of it, since even among bodies, in some cases contact is the cause of their being one, in others stickiness, or some other attribute of this sort."[16]

16. Emergence (Baas and Emmeche)

"Let $\{S_i\}_{i \in I}$ be a family of general systems or 'agents'. Let Obs^1 be observation mechanisms and Int^1 be interactions between agents.

The observation mechanism measures the properties of the agents to be used in the interactions. The interactions then generate a new kind of structure

$$S^2 = R\left(S^{1:i}, Obs^1, Int^1\right)$$

Which is the result of the interactions. This could be a stable pattern or a dynamically interacting system. We call S^2 and emergence structure which may be subject to new observational mechanisms Obs^2. This leads to [the] definition:

P is an emergent property if $P \in Obs^2(S^2)$ and $P \notin Obs^2(S^{1:i})$.

The observational mechanism may be internal or external".[17]

17. Emergence (Broad)

"... the emergent theory asserts that there are certain wholes, composed (say) of constituents A,B, and C in a relation R to each other; that all wholes composed of constituents of the same kind as A, B, and C in relations of the same kind as R have certain characteristic properties ; that A, B, and C are capable of occurring in other kinds of complex where the relation is not of the same kind as R ; and that the characteristic properties of the whole R(A, B, C) cannot, even in theory, be deduced from the most complete knowledge

---

[15] Alexander (1920), p. 45.
[16] Aristotle and Reeve (2016), p. 147.
[17] Baas and Emmeche (1997).

of the properties of A, B, and C in isolation or in other wholes which are not of the form R(A, B, C)."[18]

18. Emergence (Butterfield)

"...behaviour that is novel and robust relative to some comparison class. ...two widespread conceptions of what the comparison class is, as follows.

(1) Composites: The system is a composite; and its properties and behaviour are novel and robust compared to those of its component systems, especially its microscopic or even atomic components.
(2) Limits: The system is a limit of a sequence of systems, typically as some parameter (in the theory of the systems) goes to infinity (or some other crucial value, often zero); and its properties and behaviour are novel and robust compared to those of systems described with a finite (respectively: non-zero) parameter."[19]

19. Emergence (Churchland)

"...a property F will be said to be an emergent property (relative to [a theory] T_N) just in case.

(1) F is definitely real and instantiated;
(2) F is co-occurrent with some feature or complex circumstance recognized in T_N; but
(3) F cannot be reduced to any property postulated by or definable within T_N."[20]

20. Emergence (Darley)

"...emergent phenomena are those for which the amount of computation necessary for prediction from an optimal set of rules, classifications and analysis, even derived from an idealised perfect understanding, can never improve

---

[18] Broad (1925).
[19] Butterfield (2011).
[20] Churchland (1985).

upon the amount of computation necessary to simulate the system directly from our knowledge of the rules of its interactions."[21]

21. Emergence (Deacon)

"A term used to designate an apparently discontinuous transition from one mode of causal properties to another of a higher rank, typically associated with an increase in scale in which lower-order component interactions contribute global properties that appear irreducible to the lower-order interactions."[22]

22. Emergence (Ellis)

"Emergence E of system from its components. For example, the emergence E of nuclei out of protons and neutrons, of water or a metal or hemoglobin molecules out of the underlying nuclei and electrons, or of a human body out of its constituent cells. The issue of phase transitions is important here. These occur when a major change in the emergent state takes place, such as the transition of water from a liquid to a gaseous state when boiling occurs.

Emergence P of properties of the emergent system out of its underlying constituents once it has come into existence. How do properties of a nucleus arise out of the nature of its constituent neutrons and protons, and theirs out of the constituent quarks? How do rigidity or electrical conductivity or optical properties of a crystal, or chemical properties of a molecule, arise out of the underlying electrons, protons and neutrons? How do properties of a cell in a human body arise out of properties of its underlying biomolecules? How does behaviour arise out of those cells?".[23]

23. Emergence (Israeli and Goldenfeld)

"Emergent properties are those which arise spontaneously from the collective dynamics of a large assemblage of interacting parts."[24]

---

21 Darley (1994).
22 Deacon (2012), p. 502.
23 Ellis (2020).
24 Israeli and Goldenfeld (2006).

24. Emergence (Klee)

"Property P is emergent at a level of organization in a system, with respect to that system's lower level microstructure MS, when (and possibly only when) either.

(1) P is unpredictable in principle from MS (i.e. unpredictable even from an ideally complete theoretical knowledge of MS in the limit of scientific inquiry)

or

(2) P is novel with respect to MS

or

(3) MS exhibits a much greater degree of variance and fluctuation from moment to moment than does the level of organization where P occurs, P's constant and enduring presence in the system would or not seem to be wholly determined by MS.

(4) P has direct determinative influence and effects on at least some of the properties in MS."[25]

25. Emergence (Laughlin)

"A collective principle of organization that gives rise to a law, a relationship among measured things that is always true."[26]

26. Emergence (Lewes)

"...although each effect in the resultant of its components, the product of its factors, we cannot always trace the steps of the process, so as to see in the product the mode of operation of each factor. In this... case I propose to call the effect an emergent. It arises out of the combined agencies, but in a form which does not display the agents in action...

---

[25] Klee (1984).
[26] Laughlin (2021).

The emergent is unlike its components in so far as these are incommensurable and it cannot be reduced either to their sum or their difference. But, on the other hand, it is like its components, or more strictly speaking, it is these: nothing can be more like the coalescence of the components than the emergent which is they coalescence."[27]

27. Emergence (Mayr)

"Systems almost always have the peculiarity that the characteristics of the whole cannot (not even in theory) be deduced from the most complete knowledge of the components, taken separately or in other partial combinations. This appearance of new characteristics in wholes has been designated as emergence."[28]

28. Emergence (McLaughlin)

"If P is a property of w, then P is emergent if and only if (1) P supervenes with nomological necessity, but not with logical necessity, on properties the parts of w have taken separately or in other combinations; and (2) some of the supervenience principles linking properties of the parts of w with w's having P are fundamental laws."[29]

29. Emergence (Mill)

"The chemical combination of two substances produces, as is well known, a third substance with properties entirely different from those of either of the two substances separately, or of both of them taken together. Not a trace of the properties of hydrogen or of oxygen is observable in those of their compound, water. The taste of sugar of lead is not the sum of the tastes of its component elements, acetic acid and lead or its oxide; nor is the color of green vitriol a mixture of the colors of sulphuric acid and copper. This explains why mechanics is a deductive or demonstrative science, and chemistry not. In the one, we can compute the effects of all combinations of causes, whether real or hypothetical, from the laws which we know to govern those causes when acting separately; because they continue to observe the same laws when in combination, which they observed when separate: whatever would

---

[27] Lewes (1874), p. 368.
[28] Mayr (1982), p. 63.
[29] McLaughlin (2008).

have happened in consequence of each cause taken by itself, happens when they are together, and we have only to cast up the results. Not so in the phenomena which are the peculiar subject of the science of chemistry. There, most of the uniformities to which the causes conformed when separate, cease altogether when they are conjoined; and we are not, at least in the present state of our knowledge, able to foresee what result will follow from any new combination, until we have tried it by specific experiment.

If this be true of chemical combinations, it is still more true of those far more complex combinations of elements which constitute organized bodies; and in which those extraordinary new uniformities arise, which are called the laws of life. All organized bodies are composed of parts, similar to those composing inorganic nature, and which have even themselves existed in an inorganic state; but the phenomena of life, which result from the juxtaposition of those parts in a certain manner, bear no analogy to any of the effects which would be produced by the action of the component substances considered as mere physical agents. To whatever degree we might imagine our knowledge of the properties of the several ingredients of a living body to be extended and perfected, it is certain that no mere summing up of the separate actions of those elements will ever amount to the action of the living body itself...

...as a general rule, causes in combination produce exactly the same effects as when acting singly: but that this rule, though general, is not universal: that in some instances, at some particular points in the transition from separate to united action, the laws change, and an entirely new set of effects are either added to, or take the place of, those which arise from the separate agency of the same causes: the laws of these new effects being again susceptible of composition, to an indefinite extent, like the laws which they superseded."[30]

30. Emergence (Mandelbaum)

"...'emergent naturalism' would hold that while all entities are material in character, the varying forms of organization which matter may possess give rise to diverse properties and diverse modes of behavior, neither of which can be adequately explained, even in principle, by an appeal to any single set of laws."[31]

---

[30] Mill (1859), sec. Book III, Chap. 6.2.
[31] Mandelbaum (1974), p. 22.

31. Emergence (Morgan)

"That which becomes the stuff at the higher level of emergence is never quite what it was at the lower level from which it was derived—otherwise one would have resultants only and not emergence. Under emergent evolution there is progressive development of stuff which becomes new stuff in virtue of the higher status to which it has been raised under some supervenient kind of substantial gotogetherness."[32]

32. Emergence (Pepper)

"...emergence, which is a cumulative change, a change in which certain characteristics supervene upon other characteristics, these characteristics being adequate to explain the occurrence on their level...The theory of emergence involves three propositions: (1) that there are levels of existence defined in terms of degrees of integration; (2) that there are marks which distinguish these levels from one another over and above the degrees of integration; (3) that it is impossible to deduce the marks of a higher level from those of a lower level, and perhaps also (though this is not clear) impossible to deduce marks of a lower level from those of a higher."[33]

33. Emergence (Polanyi)

"If each higher level is to control the boundary conditions left open by the operations of the next lower level, this implies that these boundary conditions are in fact left open by the operations going on at the lower level. In other words, no level can gain control over its own boundary conditions and hence cannot bring into existence a higher level, the operations of which would consist in controlling these boundary conditions. Thus the logical structure of the hierarchy implies that a higher level can come into existence only through a process not manifest in the lower level, a process which thus qualifies as an emergence."[34]

34. Emergence (Ronald, Sipper, Capcarrère)

1. Design: The system has been constructed by the designer, by describing local elementary interactions between components (e.g.,

---

[32] Morgan (1923), pp. 192–3.
[33] Pepper (1926).
[34] Polanyi (2009).

artificial creatures and elements of the environment) in a language L_1.

2. Observation: the observer is fully aware of the design, but describes global behaviors and properties of the running system, over a period of time, using a language L_2.

3. Surprise: The language of design L_1 and the language of observation L_2 are distinct, and the causal link between the elementary interactions programmed in L_1 and the behaviors observed in L_2 is non-obvious to the observer—who therefore experiences surprise. In other words, there is a cognitive dissonance between the observer's mental image of the system's design stated in L_1 and his contemporaneous observation of the system's behavior stated in L_2."[35]

35. Emergence (Wilson)

"...the coupling of cotemporal material dependence with ontological and causal autonomy which is most basically definitive of the notion of emergence."[36]

36. Emergent Dualism (Nida-Rümelin)

"Claim 1 (Emergence of new individuals): There are specific physical conditions C such that the following holds: at any time t, if t is the time at which a particular material system M (e.g., a biological organism) first satisfies C, then with nomological necessity a subject of experience (a conscious being that belongs to an ontological category different form the one of material objects) comes into existence at t and starts at t to have M as its body.

Claim 2 (Emergence of consciousness properties): A subject cannot have consciousness properties unless the subject's body has corresponding physical properties. No change in consciousness properties is nomologically possible without a simultaneous change in corresponding physical properties of the subject's body. No two nomologically possible individuals (whether in the same world or in different worlds) can differ in their consciousness properties without a difference in the physical properties of their respective bodies."[37]

---

[35] Ronald et al. (2008), p. 291.
[36] Wilson (2021), p. 1.
[37] Nida-Rümelin (2007).

37. Explanatory Emergence (Sartenaer)

"A property E explanatorily emerges from an underlying physical basis $\{B_i\}$ [if and only if] (1) E is realized in $\{B_i\}$ [causal continuity] and (2) E downwardly acts—in Sellars' reflexive sense—on $\{B_i\}$ [explanatory discontinuity]"
　　where Sellar's sense is.
　　"... there would then exist (at least) two distinct and irreducible modes of causal determination in nature, namely (i) intra-level efficient causation that governs the succession of physical events through time, and (ii) inter-level emergent causation—other than efficient—that regulates (or harnesses, restraints, constrains, orientates, etc.) the way in which underlying intra-level causal relations unfold."[38]

38. Explanatory Emergence (Taylor)

"(Given components A, B, C... n arranged in relation r into a whole, and an observer O, property x of the whole is emergent for O iff there is no scientific explanation available to O of the fact that the following regularity obtains of natural necessity: Whenever components A, B, C...n are combined in relation r, the resulting whole instantiates property x."[39]

39. Emergence in Physics (Kivelson and Kivelson)

"An emergent behavior of a physical system is a qualitative property that can only occur in the limit that the number of microscopic constituents tends to infinity."[40]

40. Emergence Relative to a Model (Cariani)

"The emergence-relative-to-a-model view sees emergence as the deviation of the behaviours of a physical system from an observer's model of it. Emergence then involves a change in the relationship between the observer's behaviour and the physical system under observation."[41]

---

[38] Sartenaer (2016).
[39] Taylor (2015).
[40] Kivelson and Kivelson (2016).
[41] Cariani (1991).

41. Epistemological Emergence (Silberstein and McGeever)

"A property of an object or system is epistemologically emergent if the property is reducible to or determined by the intrinsic properties of the ultimate constituents of the object or system, while at the same time it is very difficult for us to explain, predict or derive the property on the basis of the ultimate constituents. Epistemologically emergent properties are novel only at a level of description. For example, even systems with very few parts and with simple mathematical rules can sometimes be said to exhibit epistemological emergence. In principle in such cases the higher-level feature, rule or law is a logical consequence of some lower-level feature, rule or law."[42]

42. Few-Many Emergence (Palacios)

"A system exhibits few-many emergence when there is a robust behavior (pattern or property) at the scale of many components that is epistemologically/ontologically novel with respect to the scale of few components."[43]

43. Fusion Emergence (Humphreys)

"… two property instances that belong to a domain D interact, and in so doing, the instances are transformed in such a way as to produce a new property instance, the key feature of which is that it does not have the original property instances as components."[44]

44. Inferential Emergence (Humphreys)

"… an entity, such as a state or a property instance, is emergent with respect to a domain D if and only if it is impossible, on the basis of a complete theory of D, to effectively predict that entity or to effectively compute a state corresponding to that feature."[45]

---

[42] Silberstein and McGeever (1999).
[43] Palacios (2022), p. 15.
[44] Humphreys (2016), p. 88.
[45] Humphreys (2008).

45. Interactive Complexity Emergence (Cunningham)

"A highly configurational and highly holistic property p is epistemically-emergent to the extent that p's proprietary entity is so interactively complex that it is difficult (or perhaps impossible) to track p's relations to the lower level properties of p's proprietary entity's components."[46]

46. Intrinsic Emergence (Crutchfield)

"…in the emergence of coordinated behavior, though, there is a closure in which the patterns that emerge are important within the system. That is, those patterns take on their 'newness' with respect to other structures in the underlying system. Since there is no external referent for novelty or pattern, we can refer to this process as 'intrinsic' emergence…. What is distinctive about intrinsic emergence is that the patterns formed confer additional functionality which supports global information processing."[47]

47. Maximal Emergence (Assad and Packard)

"Behavior is impossible to deduce from the specification [or rules generating it]."[48]

48. Measurement Emergence (Pattee)

"…I believe is the most important for evolution. I will simply call it measurement itself, but this does not help much because, as I indicated, measurement presents a fundamental problem in physics as well as biology. In classical physics, measurement is a primitive act—a pure realization that has no relation to the theory or to laws except to determine the initial conditions. However, in quantum theory measurement is an intrinsic part of the theory, so where the system being measured stops and the measuring device begins is crucial."[49]

---

[46] Cunningham (2001).
[47] Crutchfield (2008).
[48] Assad and Packard (2008).
[49] Pattee (1989).

49. Multiple Realizability Emergence (Cunningham)

"A fully configurational property p is epistemically emergent to the extent that p is so multiply and diversely realizable that p's relations to the lower level properties of p's proprietary entity's components are theoretically insignificant."[50]

50. Nominal Emergence (Bedau)

"Nominal emergence easily explains the two hallmarks of emergence. Macro-level emergent phenomena are dependent on micro-level phenomena in the straightforward sense that wholes are dependent on their constituents, and emergent phenomena are autonomous from underlying phenomena in the straightforward sense… for example, a circle consists of a collection of points, and the individual points have no shape So being a circle is a property of a 'whole' but not its constituent 'parts'—that is, it is a nominal emergent property."[51]

51. Observational Emergence (Baas and Emmeche)

"If P is an emergent property, but cannot be deduced as in [the definition of Deducible or computational emergence (Baas and Emmeche) above]."[52]

52. Ontological Emergence (Barnes)

"An entity x is dependent [if and only if] for all possible worlds w and times t at which a duplicate of x exists, that duplicate is accompanied by other concrete, contingent objects in w at t"

"An entity x is ontologically emergent iff x is fundamental and dependent."[53]

53. Ontological Emergence (Gillet)

"A property instance X, instantiated in an individual s*, is O- emergent if (i) s* is an individual which is either constituted by other individuals or is

---

[50] Cunningham (2001).

[51] Bedau (2002).

[52] Baas and Emmeche (1997).

[53] Barnes (2012).

an unconstituted, non-physical individual; (ii) X is an unrealized property instance; and (iii) X is productive and hence determinative."[54]

54. Ontological Emergence (Humphreys)

"...the ontological approach considers emergent entities to be genuinely novel features of the world itself, where an entity is ontologically emergent with respect to domain D if and only if that entity is ontologically irreducible to entities in domain D."[55]

55. Ontological Emergence (Silberstein and McGeever)

"Ontologically emergent features are neither reducible to nor determined by more basic features. Ontologically emergent features are features of systems or wholes that possess causal capacities not reducible to any of the intrinsic causal capacities of the parts nor to any of the (reducible) relations between the parts. Ontological emergence entails the failure of part–whole reductionism in both its explicit and mereological supervenience forms."[56]

56. Parts-Whole Emergence (Palacios)

"A system exhibits parts-whole emergence when the properties of the whole system are novel with respect to the properties of its parts taken in isolation."[57]

57. Representational Emergence (Sartenaer)

"A property E representationally emerges from an underlying physical basis $\{B_i\}$ [if and only if] (1) E is realized in $\{B_i\}$ [causal continuity] and (2) E downwardly acts—in Sellars' sense—on $\{B_i\}$ [representational discontinuity]."
where Sellar's sense is.
"... there would then exist (at least) two distinct and irreducible modes of causal determination in nature, namely (i) intra-level efficient causation

---

[54] Gillett (2018), Chap. 5.
[55] Humphreys (2008).
[56] Silberstein and McGeever (1999).
[57] Palacios (2022), p. 18.

that governs the succession of physical events through time, and (ii) inter-level emergent causation—other than efficient—that regulates (or harnesses, restraints, constrains, orientates, etc.) the way in which underlying intra-level causal relations unfold."[58]

58. Scale Relative Compressibility (SRC) Emergence (Pexton)

"A system is SRC-emergent if it must use higher level structural relations and terms to reduce the Kolmogorov complexity of the algorithms that represent that system (such that those algorithms only then become executable given the calculational constraints on the system)."[59]

59. Semantic Emergence (Pattee)

"… best known at the cognitive level, but could also occur at the genetic level. It is usually called creativity when it is associated with high level symbolic activity. I will call the more general concept semantic emergence… symbol systems are intrinsically rate-independent, and discrete. That is, the meaning of a gene, a sentence or a computation does not depend on how fast it is processed, and the processing is in discrete steps. At the cognitive level, we have many heuristic processes that produce semantic emergence, from simple estimation, extrapolation, and averaging, to abstraction, generalization and induction."[60]

60. Synchronic Emergence (Sartenaer)

"Synchronic emergence is an empirical relation between an emergent and its emergence basis such that (a) the emergent is constitutively determined by its emergence basis, and (b) it is not possible to trace the constitutive chain that goes from the emergence basis to the emergent."[61]

---

[58] Sartenaer (2016).
[59] Pexton (2019).
[60] Pattee (1989).
[61] Sartenaer (2015).

61. Strong Emergence (Assad and Packard)

"Behavior is deducible in theory [from the specification [or rules generating it], but its elucidation is prohibitively difficult."[62]

62. Strong Emergence (Chalmers)

"…a high level phenomenon is strongly emergent with respect to a low level domain when the high level phenomenon arises from the low level domain, truths concerning that phenomenon are not deducible even in principle from truths in the low level domain."[63]

63. Strong Emergence (Elstob)

"Let us state two conditions under which a property or entity may be said to be emergent in the strong sense that its existence cannot be explained in terms of the properties of its components considered independently of the emergent thing or level of phenomena itself.

(1) Given the existence of the emergent activity there shall be component level properties that do not exist without the presence of the phenomenal level of which the emergent entity is a member; and.

(2) The emergent entity shall have an existence that is necessarily dependent upon properties belonging only to the emergent level and not to the component level."[64]

64. Strong Emergence (Gillet)

"A property instance X, instantiated in an individual s^*, is an S- emergent property instance if (i) X is realized by property instances $P1- Pn$ of individuals $s1- sn$ (and s^* is constituted by $s1- sn$), and (ii) X is determinative."[65]

---

[62] Assad and Packard (2008).
[63] Chalmers (2006).
[64] Elstob (1984).
[65] Gillett (2018), Chap. 5.

65. Strong Metaphysical Emergence (Wilson)

"What it is for token feature S to be Strongly metaphysically emergent from token feature P on a given occasion is for it to be the case, on that occasion, (i) that S cotemporally materially depends on P, and (ii) that S has at least one token power not identical with any token power of P."[66]

66. Strong Pancomputationalist Emergence (Pexton)

"We have a system that cannot be informationally compressed at the microphysical level. AND That system finds a solution in a timescale that exceeds the calculational limits it has at the microphysical level."[67]

67. Syntactical Emergence (Pattee)

"…symmetry-breaking and chaotic dynamics…"[68]

68. Theoretical Emergence (Sartenaer)

A property E theoretically emerges from an underlying physical basis $\{B_i\}$ [if and only if] (1) E is realized in $\{B_i\}$ [causal continuity] and (2) environmental properties tend to select E through downwardly acting—in Sellars' non-reflexive and selective sense—on $\{B_i\}$ [theoretical discontinuity".[69]

69. Thermodynamic Emergence (Cariani)

"…[involves] order-from-noise, discrete macro-structures (attractors) from continuous micro processes, new structures emerge through fluctuations."[70]

70. Transformational Emergence (Humphreys)

"Transformational emergence occurs when an individual a that is considered to be a fundamental element of a domain D transforms into a different kind of individual a*, often but not always as a result of interactions with other elements of D, and thereby becomes a member of a different domain D*.

---

[66] Wilson (2021), p. 120.
[67] Pexton (2019).
[68] Pattee (1989).
[69] Sartenaer (2016).
[70] Cariani (1991).

Members of D* are of a different type from members of D. They possess at least one novel property and are subject to different laws that apply to members of D* but not to members of D."[71]

71. Weak Emergence (Assad and Packard)

"Behavior is deducible in hindsight from the specification [or rules generating it] after observing the behavior."[72]

72. Weak Emergence (Bedau)

"Macrostate P of [microstate] S with microdynamic D is weakly emergent [if and only if] P can be derived from D and S's external conditions but only by simulation."[73]

73. Weak Emergence (Chalmers)

"…a high level phenomenon is weakly emergent with respect to a low level domain when the high level phenomenon arises from the low level domain, but truths concerning that phenomenon are unexpected given the principles governing the low level domain."[74]

74. Weak Emergence (Gillet)

"A property instance X, instantiated in an individual s*, is W-emergent if (i) X is realized by property instances P1– Pn of individuals s1– sn (and s* is constituted by s1– sn), and (ii) the higher scientific law statements and/ or theories and/ or explanations taken to be true of X cannot be derived and/ or computed and/ or predicted from the lower-level scientific law statements and/ or theories and/ or explanations holding of the property instances P1– Pn that realize X in s*."[75]

---

[71] Humphreys (2016), p. 74.
[72] Assad and Packard (2008).
[73] Bedau (1997).
[74] Chalmers (2006).
[75] Gillett (2018), Chap. 5.

75. Weak Metaphysical Emergence (Wilson)

"What it is for token feature S to be Weakly metaphysically emergent from token feature P on a given occasion is for it to be the case, on that occasion, (i) that S cotemporally materially depends on P, and (ii) that S has a non-empty proper subset of the token powers had by P."[76]

A.2. Decoherence

This follows Tegmark (1993). He defines:
Coherence time $\tau = \Lambda^{-1}$.
Decoherence rate: $\Delta = \frac{\Lambda}{\lambda_{eff}^2}$ where λ_{eff}^2 is the effective wavelength of the scattering process.

So then $\tau = \Lambda^{-1} = \left(\Delta\lambda_{eff}^2\right)^{-1}$.

The following values are taken from Tables 1 and 2 in the paper.

	λ_{eff} (m)	Electron Δ (cm^{-2}s^{-1})	τ (s)	Dust Δ (cm^{-2}s^{-1})	T (s)	Bowling ball Δ (cm^{-2}s^{-1})	T (s)
Air in laboratory	10^{-11}	10^{31}		10^{-13} 10^{37}		10^{-19} 10^{45}	10^{-27}
Laboratory vacuum	10^{-11}	10^{18}		10^{0} 10^{23}		10^{-5} 10^{31}	10^{-13}
Sunlight on earth	9×10^{-7}	10^{1}		10^{7} 10^{20}		10^{-12} 10^{28}	10^{-20}
CMB	2×10^{-3}	10^{-10}		10^{12} 10^{6}		10^{-4} 10^{17}	10^{-15}

A.3. The Thermodynamic Limit

Take some property of a thermodynamic system $Q(T, V, N)$ where T is temperature, N the number of particles and V the volume. Define $n = N/V$.

[76] Wilson (2021), p. 75.

The thermodynamic limit is:

$$\lim_{N,V\to\infty} Q(T, V, N) = q(n, T)$$

The limit exists if q is finite. Since q only depends on n it is independent of the size of the system.

As an example of a system in which the limit exists, take an ideal gas. We can write its energy as:

$$E = \frac{3}{2}NkT$$

and

$$\frac{E}{V} = \frac{3}{2}nkT$$

is well-defined in the limit.

For a system where the limit doesn't exist, allow the particles to interact by gravity. Then we need to add a term for gravitational potential energy

$$P = -\frac{3GM^2}{5R}$$

If the density of the gas is constant, the mass will be proportional to the volume so

$$P \propto -\frac{V^2}{V^{\frac{1}{3}}} \propto V^{-\frac{5}{3}}$$

Then the potential energy per unit volume is given by

$$\frac{P}{V} \propto V^{-\frac{2}{3}}$$

However in the thermodynamic limit this tends to $-\infty$ i.e. the limit does not exist.

A.4. The Renormalization Group Transformation

This is an attempt at a minimal explanation of RGT which covers what I need for the exposition in the text. As such, it neglects to mention a number of key concepts. For a proper treatment, see the suggestions for further reading to Chap. 10.

Take a system made up of elements arranged on a lattice. The coupling constants between elements are {J} which in general includes interactions between neighbours, next-neighbours etc. Let the Hamiltonian of the system be

$$H = H(K, S)$$

where $K = J/T$ and S are the degrees of freedom of the system (I'm omitting the curly brackets around both).

Then assume the lattice is such that there is a coarse graining transformation which aggregates the S into blocks:

$$S' = Q(S)$$

such that

- the coarse grained degrees of freedom S' have the same range of values as S
- the new lattice has the same symmetry properties as the original

Then the Hamiltonian will have the same general form as the original, but with renormalized coupling constants given by

$$K' = R_Q(K)$$

and R_Q is the RGT, the Q subscript indicating dependence on the type of coarse graining. In general, the transformation will introduce coupling at higher orders. For example, if we start with a system with only nearest-neighbour interactions, the transformed system may include next-nearest neighbour interactions as well.

The transformation will only have fixed points in the thermodynamic limit, otherwise repeated application of the RGT will hit the system boundaries. A fixed point K* is given by

$$K^* = R_Q(K^*)$$

and will be independent of the starting value of K.

Intuitively, a fixed point means the pattern[77] of S is the same at whatever scale we view it, so there are only three possibilities.

(i) The S are all the same (corresponding to $J = \infty$ or $T = 0$)
(ii) The S are random[78] ($J = 0$ or $T = \infty$)
(iii) The S are scale invariant.

It can be shown that the first two are stable to perturbation. The third is unstable and corresponds to a phase change.

The dynamics in the vicinity of a fixed point can be investigated by linearising around the fixed point. They will also be independent of K. This is universality: the fixed points and the nearby dynamics depend only on the symmetry properties of the lattice and not on the coupling constants. Values of universal exponents can be derived from the same linearization.

A.4.1 Finite Size

To estimate the importance of finite size,[79] note that in a finite system the correlation length will be constrained by the size of the system, L. If T_C is the critical temperature, define a reduced temperature as:

$$t = \frac{T - T_C}{T_C}$$

Then assume that the correlation length ξ scales according to:

$$\xi = \xi_0 t^{-\frac{2}{3}}$$

Take the correlation length a long way from the critical point to be $\xi_0 = 10^{-9} m$. Then for a system with L = 1 cm, the correlation length hits this when $t \approx 10^{-11}$.

---

[77] More formally, does the correlation length between elements stay the same. The first and third possibilities have an infinite correlation length; the second a correlation length of zero.

[78] This is scale invariant since a combination of random variable is also a random variable.

[79] This is adapted from Goldenfeld (1992), p. 31.

A.4.2 Scale Invariance

Scale invariance means that the system looks the same at whatever scale we observe it. Take a function f which describes the system based on some length variable x. Scale invariance is when:

$$f(ax) = a^b f(x)$$

If correlations between spins follow a power law in their separation r:

$$f(r) = r^{-\nu}$$

This satisfies the definition of scale invariance since

$$f(ar) = a^{-\nu} r^{-\nu} = a^{-\nu} f(r)$$

A.5 The Ising Model

A.5.1 Basic Description

A 2×2 lattice of spins S_I which can be either up $S_i = +1$ or down $S_i = -1$. Assume interactions just among nearest-neighbours with an energy of $-J$ if the spins are aligned or $+J$ if they are in opposite directions. Define the coupling constant to be

$$K = \frac{J}{kT}$$

where T is temperature. Then the total energy of the system—the Hamiltonian—can be written:

$$H = -K \sum_{\langle i,j \rangle} S_j S_i$$

where $\langle i, j \rangle$ means the sum is restricted to neighbour pairs.

A.5.2 Analytic Results

The Onsager formula for spontaneous magnetisation:

$$M = \left(1 - (\sinh 2K)^{-4}\right)^{\frac{1}{8}}$$

The intellectual history of this expression is somewhat convoluted. Onsager apparently presented it at seminars during 1948 without proof. The first publication, again without proof is in a comment by Onsager to Rushbrooke (1949), p. 261 then proofs came in Yang (1952) and Montroll et al. (1963).

The critical temperature at which M reaches 0 will be given by:

$$K_c = \frac{1}{2} \sinh^{-1} 1 = \frac{1}{2} \ln\left(1 + \sqrt{2}\right) = 0.441$$

If J and k are normalised to unity:

$$T_c = \frac{1}{K_c} = 2.27$$

A.5.3 RGT

The full derivation of the RGT for the 2D Ising model is in Goldenfeld (1992), sec. 9.6. Here I just show a few of the key steps.

Let's say the original spins are given by σ which are transformed onto coarse grained spins S. If H is the original Hamiltonian, the coarse grained Hamiltonian H' is given by:

$$e^{H'(S_I)} = \sum_{\{\sigma_I\}} e^{H(S_I, \sigma_I)}$$

where the sum is over the original spins σ_I in the block S_I.

Note that in general the transformed Hamiltonian may involve higher-order terms. We can think of a general Hamiltonian for an Ising-type system as:

$$H = h \sum_i S_i + K_1 \sum_{\langle i,j \rangle} S_j S_i + K_2 \sum_{\langle i,j,k \rangle} S_k S_j S_i + \ldots.$$

where the first term is the effect of an external field, the second term next-neighbour interactions, the third term three-spin interactions then further terms to represent all other types of interaction. This form of Hamiltonian, along with the underlying lattice symmetry, defines a universality class.

For the basic Ising model we have $K_1 = K$ and $h = K_2 = K_3 = 0$. For the Ising model with a non-zero external field, $h \neq 0$. Other systems will have different values of the coupling constants. The RGT shows that whatever the values, the fixed points of the system and the local dynamics around these fixed points are the same: this is universality. All that matters is the form of the Hamiltonian.

The question is how to use the expression for the transformed Hamiltonian to obtain an expression for the transformation of the coupling constants. Goldenfeld does this by using perturbation theory. The result of a first-order perturbation is an approximate expression for the transformation of the coupling constants:

$$K' = 2K\Phi(K)^2$$

where

$$\Phi(K) = \frac{e^{3K} + e^{-K}}{e^{3K} + 3e^{-K}}$$

Note the recursive nature of the RGT means that this is all we need to know to solve for a fixed point. This can be done by setting $K = K' = K^*$ in the above pair of equations.

$$K^* = 2K^*\Phi(K^*)^2$$

Which implies $K^* = 0$ or $K^* = \infty$ corresponding to the low and high temperature fixed points. The third fixed point, the critical value, is given by $\Phi(K_C) = \frac{1}{\sqrt{2}}$.

Solving this expression gives $K_C \approx 0.34$ which is close to the exact value of 0.27.

The dynamics around the fixed point are given by an eigenvalue of the transformation. The first-order approximation gives an eigenvalue of 1.62; again quite close to the exact value of 1.73.

A.5.4 An External Field

It is easy to modify the simulation code to take account of an external magnetic field.[80] However if computing resources are limited the effective field approach is much simpler. In the case of a single spin, take the case in which the energies of the states are h and $-h$, where h is the strength of the external field. Then the probabilities of being in each state will be:

$$p(+1) = \frac{e^h}{e^h + e^{-h}}; \; p(-1) = \frac{e^{-h}}{e^h + e^{-h}}$$

and the expected value will be:

$$\langle s \rangle = 1\frac{e^h}{e^h + e^{-h}} + (-1)\frac{e^{-h}}{e^h + e^{-h}} = \tanh(h)$$

For more than one spin, assume each spin is affected both by the external field and an effective field resulting from the action of all the other spins. Then we can write the total field as:

$$h^{eff} = h + h'$$

where h′ represents the magnetic field due to all the spins in the lattice. The effective field will be in the same direction as the external field so will tend to reinforce its effect. The more spins in the system, the stronger this effective field will be and the steeper will be the magnetisation curve.

This is the Weiss model of ferromagnetism. To solve the model, to assume the additional field is proportional to the external field then show that the assumption is consistent. For a clear treatment, see Goldenfeld (1992), Sec. 3.7.1.

---

[80] The Octave code available at www.TheMaterialWorld.net allows non-zero external fields.

References

Alexander S (1920) Space, Time, and Deity, Vol 2. Macmillan, London

Aristotle, Reeve CDC (2016) Metaphysics. Hackett Publishing Company, Indianapolis; Cambridge

Assad A, Packard NH (2008) Emergence. In: Bedau MA, Humphreys P (eds) Emergence: Contemporary Readings in Philosophy and Science. The MIT Press

Baas NA, Emmeche C (1997) On emergence and explanation. Intellectica 25:67–83. https://doi.org/10.3406/intel.1997.1558

Barnes E (2012) Emergence and Fundamentality. Mind 121:873–901. https://doi.org/10.1093/mind/fzt001

Bedau MA (1997) Weak Emergence. Philosophical Perspectives 11:375–399. https://doi.org/10.1111/0029-4624.31.s11.17

Bedau MA (2002) Downward Causation and the Autonomy of Weak Emergence. Principia: An International Journal of Epistemology 06:5–50. https://periodicos.ufsc.br/index.php/principia/article/view/17003

Bishop RC, Silberstein M, Pexton M (2022) Emergence in context: a treatise in twenty-first century natural philosophy. Oxford University Press, Oxford, United Kingdom

Broad CD (1925) The Mind and Its Place in Nature. Harcourt, Brace and Company, New York

Butterfield J (2011) Less is Different: Emergence and Reduction Reconciled. Found Phys 41:1065–1135. https://doi.org/10.1007/s10701-010-9516-1

Cariani P (1991) Emergence and artificial life. In: Langton CG, Taylor CE, Farmer JD, Rasmussen S (eds) Artificial Life II. Addison-Wesley, Reading, MA

Chalmers DJ (2006) Strong and Weak Emergence. In: Davies P, Clayton P (eds) The Re-Emergence of Emergence: The Emergentist Hypothesis From Science to Religion. Oxford University Press

Churchland PM (1985) Reduction, Qualia, and the Direct Introspection of Brain States. The Journal of Philosophy 82:8. https://doi.org/10.2307/2026509

Crutchfield JP (2008) Is Anything Ever New? Considering Emergence. In: Bedau MA, Humphreys P (eds) Emergence: Contemporary Readings in Philosophy and Science. The MIT Press.

Cunningham B (2001) The Reemergence of 'Emergence.' Philos of Sci 68:S62–S75. https://doi.org/10.1086/392898

Darley V (1994) Emergent Phenomena and Complexity. In: Brooks RA, Maes P (eds) Artificial Life IV. The MIT Press, pp 407–412

Deacon TW (2012) Incomplete nature: how mind emerged from matter, 1st ed. W.W. Norton & Co, New York

Ellis G (2020) Emergence in Solid State Physics and Biology. Found Phys 50:1098–1139. https://doi.org/10.1007/s10701-020-00367-z

Elstob M (1984) Emergentism and Mind. In: Trappl R (ed) Cybernetics and Systems Research 2. North-Holland

Gillett C (2018) Reduction and emergence in science and philosophy. Cambridge University Press, Cambridge

Goldenfeld N (1992) Lectures on phase transitions and the renormalization group. Addison-Wesley, Reading (Mass.)

Guay A, Sartenaer O (2016) A new look at emergence. Or when after is different. Euro Jnl Phil Sci 6:297–322. https://doi.org/10.1007/s13194-016-0140-6

Humphreys P (2016) Emergence. Oxford University Press, New York, NY, United States of America

Humphreys P (2008) Computational and Conceptual Emergence. Philos of Sci 75:584–594. https://doi.org/10.1086/596776

Israeli N, Goldenfeld N (2006) Coarse-graining of cellular automata, emergence, and the predictability of complex systems. Phys Rev E 73:026203. https://doi.org/10.1103/PhysRevE.73.026203

Kivelson S, Kivelson SA (2016) Defining emergence in physics. npj Quant Mater 1:16024. https://doi.org/10.1038/npjquantmats.2016.24

Klee RL (1984) Micro-Determinism and Concepts of Emergence. Philos of Sci 51:44–63. https://doi.org/10.1086/289163

Laughlin RB (2021) Is Emergence Fundamental? https://www.youtube.com/watch?v=qT9iDcajqMo. Accessed 11 Jun 2024

Lewes GH (1874) Problems of life and mind, first series, The Foundations of a Creed, Vol 2. Osgood, Boston

Mandelbaum M (1974) History, man, & reason: a study in nineteenth-century thought. Johns Hopkins Press, Baltimore

Mayr E (1982) The growth of biological thought: diversity, evolution, and inheritance. Belknap Press, Cambridge, Mass.

McLaughlin BP (2008) Emergence and Supervenience. In: Bedau MA, Humphreys P (eds) Emergence: Contemporary Readings in Philosophy and Science. MIT Press, Cambridge

Mill JS (1859) System of Logic, Rationative and Inductive; Being a Connected View of The Principles of Evidence and the Methods of Scientific Investigation. Harper and Brothers, New York

Montroll EW, Potts RB, Ward JC (1963) Correlations and Spontaneous Magnetization of the Two-Dimensional Ising Model. Journal of Mathematical Physics 4:308–322. https://doi.org/10.1063/1.1703955

Morgan CL (1923) Emergent Evolution. Williams and Norgate, London

Nida-Rümelin M (2007) Dualist Emergentism. In: McLaughlin BP, Cohen JD (eds) Contemporary debates in philosophy of mind. Blackwell Pub, Malden, Mass

Palacios P (2022) Emergence and reduction in physics. Cambridge University Press, Cambridge,

Pattee HH (1989) Simulations, realizations, and theories of Life. In: Langton C (ed) Artificial Life. Addison-Wesley, Redwood City, CA

Pepper SC (1926) Emergence. The Journal of Philosophy 23:241. https://doi.org/10.2307/2014779

Pexton M (2019) Computational Emergence : Weak and strong. In: Gibb SC (ed) The Routledge handbook of emergence. Routledge, New York

Polanyi M (2009) The tacit dimension. University of Chicago Press, Chicago

Ronald EMA, Sipper M, Capcarrère MS (2008) Design, Observation, Surprise! A Test of Emergence. In: Bedau MA, Humphreys P (eds) Emergence: Contemporary Readings in Philosophy and Science. The MIT Press, Cambridge

Rueger A (2000) Physical Emergence, Diachronic And Synchronic. Synthese 124:297–322. https://doi.org/10.1023/A:1005249907425

Rushbrooke GS (1949) On the theory of regular solutions. Nuovo Cim 6:251–263. https://doi.org/10.1007/BF02780989

Sartenaer O (2016) Sixteen Years Later: Making Sense of Emergence (Again). J Gen Philos Sci 47:79–103. https://doi.org/10.1007/s10838-015-9312-x

Sartenaer O (2015) Synchronic vs. diachronic emergence: a reappraisal. Euro Jnl Phil Sci 5:31–54. https://doi.org/10.1007/s13194-014-0097-2

Searle J (2008) Reductionism and the Irreducibility of Consciousness. In: Bedau MA, Humphreys P (eds) Emergence: Contemporary Readings in Philosophy and Science. The MIT Press, Cambridge

Silberstein M, McGeever J (1999) The Search for Ontological Emergence. Philosophical Quarterly 49:201–214. https://doi.org/10.1111/1467-9213.00136

Stephan A (2006) The dual role of 'emergence' in the philosophy of mind and in cognitive science. Synthese 151:485–498. https://doi.org/10.1007/s11229-006-9019-y

Taylor E (2015) An explication of emergence. Philos Stud 172:653–669. https://doi.org/10.1007/s11098-014-0324-x

Tegmark M (1993) Apparent wave function collapse caused by scattering. Found Phys Lett 6:571–590. https://doi.org/10.1007/BF00662807

Wilson JM (2021) Metaphysical Emergence. Oxford University Press, Oxford

Winning J, Bechtel W (2019) Being emergence vs. pattern emergence. In: Gibb SC (ed) The Routledge handbook of emergence. Routledge, New York

Yang CN (1952) The Spontaneous Magnetization of a Two-Dimensional Ising Model. Phys Rev 85:808–816. https://doi.org/10.1103/PhysRev.85.808

Index

Printed in the United States
by Baker & Taylor Publisher Services